Transmission Electron Microscopy in Micro-nanoelectronics

Transmission Electron Microscopy in Micro-nanoelectronics

Edited by
Alain Claverie

Series Editor
Mireille Mouis

First published 2013 in Great Britain and the United States by ISTE Ltd and John Wiley & Sons, Inc.

Apart from any fair dealing for the purposes of research or private study, or criticism or review, as permitted under the Copyright, Designs and Patents Act 1988, this publication may only be reproduced, stored or transmitted, in any form or by any means, with the prior permission in writing of the publishers, or in the case of reprographic reproduction in accordance with the terms and licenses issued by the CLA. Enquiries concerning reproduction outside these terms should be sent to the publishers at the undermentioned address:

ISTE Ltd
27-37 St George's Road
London SW19 4EU
UK

www.iste.co.uk

John Wiley & Sons, Inc.
111 River Street
Hoboken, NJ 07030
USA

www.wiley.com

© ISTE Ltd 2013

The rights of Alain Claverie to be identified as the author of this work have been asserted by him in accordance with the Copyright, Designs and Patents Act 1988.

Library of Congress Control Number: 2012952185

British Library Cataloguing-in-Publication Data
A CIP record for this book is available from the British Library
ISBN: 978-1-84821-367-8

Printed and bound in Great Britain by CPI Group (UK) Ltd, Croydon, Surrey CR0 4YY

Table of Contents

Introduction.. xi

**Chapter 1. Active Dopant Profiling in the TEM by
Off-Axis Electron Holography** 1
David COOPER

   1.1. Introduction.. 1
   1.2. The Basics: from electron waves to phase images 3
      1.2.1. Electron holography for the measurement of electromagnetic fields 3
      1.2.2. The electron source.. 6
      1.2.3. Forming electron holograms using an electron biprism 6
      1.2.4. Care of the electron biprism.............................. 10
      1.2.5. Recording electron holograms 11
      1.2.6. Hologram reconstruction 12
      1.2.7. Phase Jumps... 15
   1.3. Experimental electron holography 16
      1.3.1. Fringe contrast, sampling and phase sensitivity 16
      1.3.2. Optimizing the beam settings for an electron holography experiment.. 20
      1.3.3. Optimizing the field of view using free lens control 21
      1.3.4. Energy filtering for electron holography 24
      1.3.5. Minimizing diffraction contrast 25
      1.3.6. Measurement of the specimen thickness 26
      1.3.7. Specimen preparation 28
      1.3.8. The electrically inactive thickness 30
   1.4. Conclusion 33
   1.5. Bibliography 33

Chapter 2. Dopant Distribution Quantitative Analysis Using STEM-EELS/EDX Spectroscopy Techniques 37
Roland PANTEL and Germain SERVANTON

2.1. Introduction... 37
   2.1.1. Dopant analysis challenges in the silicon industry 37
   2.1.2. The different dopant quantification and imaging methods...... 38
2.2. STEM-EELS-EDX experimental challenges for
quantitative dopant distribution analysis 41
   2.2.1. Instrumentation present state-of-the-art and
   future challenges 41
2.3. Experimental conditions for STEM spectroscopy
impurity detection.. 43
   2.3.1. Radiation damages 43
   2.3.2. Particularities of EELS and EDX spectroscopy techniques 44
   2.3.3. Equipments used for the STEM-EELS-EDX analyses
   presented in this chapter................................. 49
2.4. STEM EELS-EDX quantification of dopant distribution
application examples 49
   2.4.1. EELS application analysis examples 49
   2.4.2. EDX application analysis examples 54
2.5. Discussion on the characteristics of STEM-EELS/EDX and
data processing 59
2.6. Bibliography 59

Chapter 3. Quantitative Strain Measurement in Advanced Devices: A Comparison Between Convergent Beam Electron Diffraction and Nanobeam Diffraction 65
Laurent CLÉMENT and Dominique DELILLE

3.1. Introduction.. 65
3.2 Electron diffraction technique in TEM (CBED and NBD) 66
   3.2.1. CBED patterns acquisition and analysis 66
   3.2.2. NBD patterns acquisition and analysis 70
3.3. Experimental details..................................... 71
   3.3.1. Instrumentation and setup............................ 71
   3.3.2. Samples description 72
3.4. Results and discussion 72
   3.4.1. Strain evaluation in a pMOS transistor integrating eSiGe
   source and drain – a comparison of CBED and NBD techniques 72
   3.4.2. Quantitative strain measurement in advanced
   devices by NBD 75
3.5. Conclusion 78
3.6. Bibliography 78

Chapter 4. Dark-Field Electron Holography for Strain Mapping 81
Martin HŸTCH, Florent HOUDELLIER, Nikolay CHERKASHIN, Shay REBOH,
Elsa JAVON, Patrick BENZO, Christophe GATEL, Etienne SNOECK and
Alain CLAVERIE

    4.1. Introduction. 81
    4.2. Setup for dark-field electron holography 83
    4.3. Experimental requirements. 85
    4.4. Strained silicon transistors with recessed sources
    and drains stressors 87
        4.4.1. Strained silicon p-MOSFET 87
    4.5. Thin film effect 92
    4.6. Silicon implanted with hydrogen 93
    4.7. Strained silicon n-MOSFET 94
    4.8. Understanding strain engineering. 96
    4.9. Strained silicon devices relying on stressor layers 97
    4.10. 28-nm technology node MOSFETs 99
    4.11. FinFET device 101
    4.12. Conclusions 103
    4.13. Bibliography 103

Chapter 5. Magnetic Mapping Using Electron Holography 107
Etienne SNOECK and Christophe GATEL

    5.1. Introduction. 107
    5.2. Experimental 108
        5.2.1. The Lorentz mode 110
        5.2.2 The "ϕ^E" problem. 111
    5.3. Hologram analysis: from the phase images to the
    magnetic properties. 118
        5.3.1. The simplest case: homogeneous specimen
        of constant thickness 119
        5.3.2. The general case 122
    5.4. Resolutions 124
        5.4.1. Magnetic measurements accuracy 124
        5.4.2. Spatial resolution 126
    5.5. One example: FePd (L10) epitaxial thin film exhibiting a
    perpendicular magnetic anisotropy (PMA) 126
    5.6. Prospective and new developments 130
        5.6.1. Enhanced signal and resolution 130
        5.6.2. In-situ switching 131
    5.7. Conclusions. 132
    5.8. Bibliography 133

Chapter 6. Interdiffusion and Chemical Reaction at Interfaces by TEM/EELS 135
Sylvie SCHAMM-CHARDON

- 6.1. Introduction. 135
- 6.2. Importance of interfaces in MOSFETs. 135
- 6.3. TEM and EELS 137
- 6.4. TEM/EELS and study of interdiffusion/chemical reaction at interfaces in microelectronics 137
  - 6.4.1. Thickness measurement 138
  - 6.4.2. Atomic structure analysis 139
  - 6.4.3. EELS analysis. 141
  - 6.4.4. Sample preparation 143
- 6.5. HRTEM/EELS as a support to developments of RE- and TM-based HK thin films on Si and Ge 144
  - 6.5.1. Introduction 144
  - 6.5.2. HRTEM/EELS methodology 145
  - 6.5.3. Illustrations 154
- 6.6. Conclusion 158
- 6.7 Bibliography 158

Chapter 7. Characterization of Process-Induced Defects. 165
Nikolay CHERKASHIN and Alain CLAVERIE.

- 7.1. Interfacial dislocations 166
  - 7.1.1. Si(100)/Si(100) direct wafer bonding (DWB) 167
  - 7.1.2. SiGe heterostructures 170
- 7.2. Ion implantation induced defects 172
  - 7.2.1. Defects of interstitial type 173
  - 7.2.2. Defects of vacancy type 187
- 7.3. Conclusions. 193
- 7.4. Bibliography 193

Chapter 8. *In Situ* Characterization Methods in Transmission Electron Microscopy 199
Aurélien MASSEBOEUF

- 8.1. Introduction. 199
- 8.2. *In situ* in a TEM 200
  - 8.2.1. Temperature control and irradiation 201
  - 8.2.2. Electromagnetic field 201
  - 8.2.3. Mechanical. 202
  - 8.2.4. Chemistry 202
  - 8.2.5. Light 203
  - 8.2.6. Multiple and movable currents 203

8.3. Biasing in a conventional TEM 204
    8.3.1. Multiple contacts 204
    8.3.2. Movable contacts 206
    8.3.3. Comparison 206
8.4. Sample design 208
    8.4.1. Focused ion beam. 208
    8.4.2. TEM windows. 209
8.5. Conclusions. 211
8.6. Bibliography. 211

Chapter 9. Specimen Preparation for Semiconductor Analysis. 219
David COOPER and Gérard BEN ASSAYAG

9.1. The focused ion beam tool 220
9.2. Ion-sample interaction 221
9.3. Beam currents and energies for specimen preparation. 225
9.4. Practical specimen preparation 228
9.5. In situ lift-out. 228
9.6. H-bar technique 232
9.7. Broad beam ion milling. 233
9.8. Mechanical wedge polishing. 235
9.9. Conclusion 235
9.10 Biblioraphy. 236

List of Authors 237

Index 241

Introduction

The MOS (Metal Oxyde Semiconductor) transistor is the key component driving the electronic logic revolution for the past 50 years ever since what has become known as Moore's law was first published [MOR 65]. Moore claimed that the number of components inside a single chip would rise exponentially, increasing by a factor of two every year and a half. After 50 years and 30 technology nodes, and despite the fact that some physicists had predicted a real MOS limit for 50 nm gate lengths and below, Moore's law still does not show any inflexion. The transistor gate length has continued to decrease from a few microns to a few tens of nanometers and the number of components per chip has crossed over the billions. This trend continues at a constant speed, respecting the initial Moore's law. Why then are the limitations predicted in the literature still not observed? First, these limitations were based on the idea that evolution was only a matter of scaling and that ultimate transistors would look like the old transistors, that is planar, mostly made up of conventional Si and SiO_2 and fabricated using basically the same processes of that in the 1980s. In fact, transistors still evolve because new materials are being integrated; they are built following new architectural rules and fabricated using different, alternative, processes.

Although the "scaling down" evolution was accompanied, and sometimes even guided, by process simulations that were based on robust, well-understood and physics-based modeling, today's evolution is more complex, sometimes looking erratic, and involves exotic materials of uncertain physical and chemical characteristics, packed together using processes in which thermodynamics plays at best a second role. More than ever before, it is necessary to experimentally access the exact chemical composition and the crystalline state of these components with an extraordinary sensitivity and at nanometer resolution. This is the prerequisite condition, which is used not only to validate technological options, but most importantly to invent and calibrate the new generation of process simulators that will again be needed to continue the incredible adventure of microelectronics.

At the same time, transmission electron microscopy (TEM) is experiencing a revolution. For many years, TEM, seen by some as the last avatar of the optical microscope, had pursued the dream of "seeing the atoms", concentrating mostly on improving the spatial resolution. Well, this is done, and no one doubts that high-resolution TEM did help considerably in figuring out how materials are made. However, today's availability of highly coherent electron sources, sensitive detectors, imaging filters and particularly aberration correctors has radically changed the type and quality of the information that can be obtained by TEM.

The first revolution comes from the "possibility to image fields" using electron holography. The possibility of quantifying, and mapping, electrostatic fields within a device is a smart answer to the everlasting question of "where are the active dopant atoms?" Mapping the strain fields introduced in the channel of a device to boost carrier mobility is mandatory to understand and optimize its performance. Moreover, the combination of intense nanoprobes and sensitive detectors can be used to dose impurity contents and identify chemical compounds that may form, intentionally or not, in the course of processing.

This book aims to present in a simple and practical way the new quantitative techniques based on TEM that have been recently invented or developed to address most of the challenging issues scientists and process engineers face to characterize or optimize semiconductor layers and devices. Several of these techniques are based on electron holography; others take advantage of the possibility to focus on intense beams within nanoprobes. Strain measurements and mappings, dopant activation and segregation, interfacial reactions at the nanoscale, defects identification, in situ experiments and specimen preparation by Focused Ion Beam (FIB) are among the topics presented in this book. After a brief presentation of the underlying theory, each technique is illustrated through examples from the lab or from the fab.

TEMs are now present in large numbers not only in academic but also in industrial research centers and fabrication plants. Some of the techniques introduced above and extensively described in the following chapters are not widespread, sometimes suffering from the *a priori* statement that they are "difficult". We believe that it is not the case and hope to convince every reader, scientist or engineer to set up and use these techniques in his or her own environment taking advantage of the "existing" or "to be bought soon" equipment.

The authors of this book have lots of experience in characterizing "real" devices, answering materials science questions arising when trying to accompany, sometimes guide, technological developments aimed at rendering electronic devices smaller, faster and cheaper while consuming less energy. This experience has been gained through daily work in public (CNRS and CEA) or private (STMicroelectronics) laboratories, often collaborating together within projects or networks financially supported by several institutions among which we want to cite the European Commission (FP6 then FP7 programs), the French ANR (White and R2N programs) and MINEFI (Alliance Nano2012) and the CNRS (METSA Network). We sincerely thank all of them for their support and help in developing and installing TEM as the indispensable companion tool of research and industry along the nanoelectronics pathway.

Bibliography

[MOR 65] MOORE G. E., "Cramming more components onto integrated circuits", *Electronics*, vol. 38, no. 8, pp. 114–117, April 1965.

Chapter 1

Active Dopant Profiling in the TEM by Off-Axis Electron Holography

1.1. Introduction

Electron holography is a powerful transmission electron microscopy (TEM)-based technique that can be used to measure the phase change of an electron wave that has passed through a region of interest compared to the phase of an electron wave that has passed through only a vacuum. As the phase of an electron is sensitive to the magnetic, electrostatic and strain fields that can be found in and around a specimen, electron holography is a unique method that can be used to recover all of these properties with nanometer-scale resolution. The electrostatic potential in semiconductor materials is modified by the presence of active dopants. At this time, when only a few dopant atoms can affect the properties of an electronic device, electron holography provides a unique opportunity to look inside these devices and to learn about the activity of the dopant atoms. Characterization techniques such as secondary ion mass spectrometry and atom probe tomography cannot differentiate between active and inactive dopants. Other techniques such as scanning capacitance microscopy and scanning spreading resistance microscopy, which are capable of measuring the active dopants at the surface of specimens, may well have problems adapting to the latest generations of semiconductor materials that can consist of doped nanowires and three-dimensional structures. Therefore, electron holography is unique in that it allows the position of active dopants to be measured inside a specimen with 1 nm spatial resolution today [COO 11], and potentially atomic resolution in the future.

Chapter written by David COOPER.

It was Gabor who introduced electron holography in his paper "Microscopy by Reconstructed Wavefronts" in 1948 [GAB 48]. Gabor realized that the measurement of the phase of an electron beam would allow the aberrations of an optical system to be eliminated. These ideas have been used in what is now known as high-resolution electron holography that have provided the first examples of sub-Ångström imaging [ORC 95]. Today, electron holography is used to describe any method that allows both the amplitude and phase information that is contained in an electron wave to be reconstructed. There are many different methods for performing electron holography, notably in-line holography that has been successfully used for the characterization of strain, dopant and magnetic fields. However, it is off-axis electron holography that is the most widely used. For simplicity, from now on, it will be referred to as electron holography. Here, a Mollenstedt–Duker biprism is used; this is a charged wire, normally located in the selected area aperture plane in a microscope. The biprism is used to tilt a reference wave so that it interferes with an object wave to provide an interference pattern in the image plane. From this interference pattern, which is also known as the electron hologram, the phase of the electron wave can be reconstructed. It was not until the 1980s when groups led by Tonomura, Pozzi and Lichte began to successfully use electron holography to solve materials science problems. However, the invention of stable and coherent electron sources in the 1990s finally allowed electron holography to become more widespread. Indeed, using the latest, ultrastable electron microscopes in 2012, electron holography has become a much more user-friendly technique that provides the microscopist with wonderful opportunities to solve materials science problems that are not available elsewhere.

This chapter is designed to show the reader how to perform electron holography in a transmission electron microscope and then how to use electron holography for dopant profiling. There are many books and reviews that deal with the theory and background in detail that should be consulted for a more complete discussion of the aspects discussed here. This chapter is designed to provide a "hands-on" approach regarding electron holography that will allow the readers to be able to get the most out of their microscope and avoid many of the common and not-so-well-known problems that can be encountered when performing electron holography.

Experimental results have been used to illustrate everything that is discussed here. The experimental conditions have been kept as constant as possible. All examples shown here were acquired using an FEI Titan TEM operated at 200 kV. Unless otherwise discussed, the Lorentz lens was used with the conventional objective lens switched off. Although the microscope used here has a probe corrector, it was not used. The presence of the probe corrector meant that the third conventional lens was switched off in order to be able to achieve the astigmatism that is required for electron holography. For recording the electron holograms, a charge-coupled device (CCD) camera attached to a Gatan energy filter was used.

This provides convenience as the image is observed at a low magnification on the TEM viewing screen to allow the whole of the sample, beam and biprism to be observed at the same time with the additional magnification then provided in the energy filter. In addition, the energy filter can be used to improve the hologram contrast. Unless otherwise stated, a 2,048 × 2,048 pixel CCD camera was used in "double binning" mode to provide 1,024 × 1,024 pixel images. Although the examples shown were acquired using a Titan TEM, everything discussed in this chapter can be transferred to any other type of TEM that is equipped with an electron biprism in the selected area plane.

1.2. The Basics: from electron waves to phase images

1.2.1. *Electron holography for the measurement of electromagnetic fields*

The phase of an electron wave that has passed through a specimen will be changed by the electromagnetic field. This phase change is given by:

$$\phi(x) = C_E \int V(x,z)\,dz - \frac{e}{\hbar} \iint B_\perp(x,z)\,dxdz$$

where z is the direction of the incident electron beam, x is the direction in the plane of the specimen, V_0 is the electrostatic potential and $B_\perp$ is the component of the magnetic induction that is perpendicular to both x and z [TON 87]. When examining specimens containing dopants, it is assumed that there is no magnetic field present. For the measurement of electrostatic potentials, the interaction constant, C_E is given by:

$$C_E = \frac{2\pi}{\lambda} \frac{E_k + E_0}{E_k(E_k + 2E_0)}$$

where λ is the electron wavelength, E_0 is the rest energy of the electron and E_k is the kinetic energy of the electron. The interaction constant is 7.29×10^6 rads $V^{-1}m^{-1}$ for 200 kV electrons and 6.53×10^6 rads $V^{-1}m^{-1}$ for 300 kV electrons. Figure 1.1 shows C_E plotted for a range of microscope operating voltages revealing that the incident electrons interact more with the electrostatic potential at lower energies.

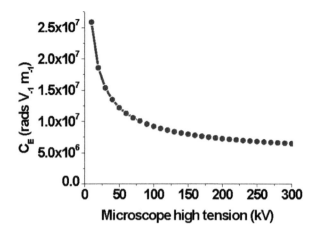

Figure 1.1. C_E as a function of the energy of the electron beam.

Following the notation of Hytch, when understanding the origin of the different phases that are measured by electron holography, we can write the phase as having four different components [HYT 11].

$$\phi_g(r) = \phi_g^G(r) + \phi_g^C(r) + \phi_g^M(r) + \phi_g^E(r)$$

where ϕ^G refers to the geometric phase that describes the distortion from the crystal lattice, ϕ^C refers to the crystalline phase resulting from the scattering of electrons from the crystal potential, ϕ^M is the magnetic contribution and ϕ^E is the contribution from the electrical fields in and around the specimen. For the purpose of this chapter, which concentrates on dopant profiling by electron holography, we will assume that the specimen is both non-magnetic and has been tilted to a weakly diffracting orientation and will only be concerned with the term $\phi_g^E(r)$. Within this term, the measured phase will have two components, the mean inner potential (MIP) V_0 and the dopant-related potential V_E.

$$V^E(r) = V_0(r) + V_E(r)$$

The MIP is defined as the volume average of the electrostatic potential in a specimen. The MIP can be calculated by using a non-binding approximation, which considers the sample as an array of neutral atoms and gives an upper limit, as it does not account for the distribution of valence electrons due to bonding. The electron scattering factors, f_{el} for each atom in the volume, Ω can be summed over the unit cell [REI 89].

$$V_0 = \left(\frac{h^2}{2\pi me\Omega}\right)\sum_{\Omega} f_{el}(0)$$

For example, to calculate the MIP in GaAs, the scattering factors can be looked up and are 7.143 nm^{-1} for Ga and 7.3686 nm^{-1} for As [REZ 94]. Given that there are four of each type of atom in a unit cell with a lattice parameter 0.5653 nm, the MIP can easily be calculated.

$$\frac{6.626\times10^{-34}}{2\pi\times9.109\times10^{-31}1.602\times10^{-19}}\times\frac{(4\times7.143\times10^{-9})+(4\times7.369\times10^{-9})}{(0.5653\times10^{-9})^3}=15.4\ V$$

A lower limit for the MIP can also be calculated by adding the contribution of the valence elections to the non-binding approximation [BET 28, RAD 70].

The MIP can be measured directly by electron holography, and values that have been measured experimentally for well-known materials are 11.9 V for Si, 10.1 V for SiO$_2$ and 14.5 V for GaAs [KRU 06]. What is important for dopant profiling is that the values of mean inner potential present in most semiconductors are at least an order of magnitude larger than expected in an electrical junction. Thus, very small changes in specimen thickness can completely mask the potential measured due to the presence of active dopants.

The MIP is sensitive to the charge density distributions that are related to the bonding and ionicity in a crystal. As such, electron holography has been successfully used to visualize these polarization fields in a range of materials, for example for nitrides, such as GaN, InGaN and AlGaN, and also in ferroelectric perovskites. For more detail on this subject, the work of the groups in Arizona led by Molly McCartney [MCC 07] and in Dresden led by Hannes Lichte [LIC 07] should be referenced.

As well as the MIP, the active dopants present in the material will additionally contribute toward the measured phase change. Therefore, in a homogeneous material of constant thickness, if sufficient care is taken, the phase image will be directly related to the distribution of active dopant atoms.

1.2.2. The electron source

For performing off-axis electron holography, a stable electron microscope equipped with a field emission gun (FEG) is required. The FEG provides a bright, stable, coherent source of electrons to form an interference pattern. The spatial coherence is described by the size of the emitting area, and the smaller this area, the brighter the source. The brightness, β, is defined as the current density per unit solid angle of the source, and it can be measured if the emission current i_e, the diameter of the focused beam d_0 and the convergence semi-angle of the beam from the condenser aperture α_0 are known [ALL 99].

$$\beta = \frac{4i_e}{(\pi d_0 \alpha_0)^2}$$

The contrast, μ, of the interference fringes can be related to the brightness of the source where I_{coh} corresponds to the coherent current and λ, the wavelength of the incident electrons.

$$I_{coh} = \frac{-\beta ln\mu}{2\pi / \lambda^2}$$

Thus, the brightness of the source is directly proportional to the coherent current. A brighter source will improve an electron microscope's ability to form higher contrast holograms, thus improving the signal to noise ratio for an identical hologram acquisition period.

Temporal coherence is dependent on the natural energy spread of the electron beam and on fluctuations in the beam energy caused by noise in the high-tension voltage supply and objective lens current. Temporal coherence can be improved using low-noise power supplies and a well-designed room environment. However, in most modern TEMs, for medium resolution electron holography, it is spatial coherence that dominates the hologram forming properties of the microscope. Thus, the interference fringes can be optimized by choosing beam settings that use electrons emitted from the very center of the electron source.

1.2.3. Forming electron holograms using an electron biprism

Figure 1.2 shows a schematic for electron holography. A coherent electron wave passes through a sample and is interfered with by an electron wave that has passed through only a vacuum by using an electron biprism. The biprism acts to create two virtual sources, S1 and S2. These virtual sources can be thought of as being comparable to Young's slits experiment.

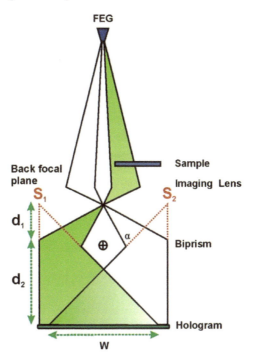

Figure 1.2. *Schematic diagram showing the effect of the biprism in forming an electron hologram*

Increasing the biprism voltage will increase the width of the interference pattern by moving the sources further apart. However, the limited coherence of the source places will affect the contrast of the interference fringes, as the electrons must then travel a greater distance to form the interference pattern. The width of the hologram, W, is:

$$W = 2\left(\frac{d_1 + d_2}{d_1}\right)\left(\alpha \frac{d_1 d_2}{d_1 + d_2} - R\right)$$

where α is the deflection angle of the electron wave due to the voltage applied to the biprism and R is the radius of the biprism [MIS 81]. From the equation, it is clear that the biprism should be as narrow as possible. The voltage that must be applied to the wire to achieve the overlap between the two parts of the electron wave on either side of it is directly proportional to the diameter of the wire. However, the increasing separation of the virtual sources required to push the biprism shadow out of the field of view will reduce the contrast quality in the hologram due to the limited coherence of the source. The biprism edge will also lead to Fresnel diffraction at the edges of the hologram, but this can be removed from the field of view by increasing the applied voltage, again at the expense of fringe contrast. Figure 1.3 shows the formation of an electron hologram at different biprism voltages. At zero volts, the biprism can be seen in the center of the pattern and Fresnel fringes are clearly observed coming from each side of the biprism. As the biprism voltage is increased, the waves begin to overlap and hologram fringes are formed. At higher voltages, finer fringes and wider patterns are observed. For the hologram formed using a biprism voltage of 50 V, several regular and fine fringes can be seen. The fringes in the center of the pattern can contain information about the amplitude and phase shift of the electron wave that has passed through the region of interest. In very simple terms, information about the phase change of the electrons can be determined from the shifts of the position of the interference fringes. The amplitude can be determined from the changes in the intensity of the fringes.

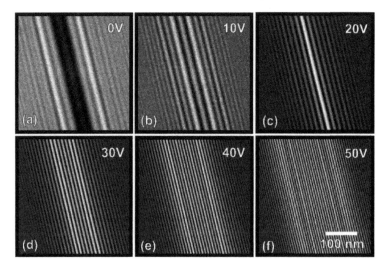

Figure 1.3. *Electron holograms acquired using different voltages applied to the biprism*

The interference fringe spacing, s, provides the spatial resolution obtainable from a hologram and can be calculated by

$$s = \lambda \left(\frac{d_1 + d_2}{2\alpha d_1} \right)$$

By increasing the voltage on the biprism, the width of the interference pattern will increase and the fringe spacing will decrease. Figures 1.4(a) and (b) show experimentally measured values of the width and fringe spacings for a range of biprism voltages that may be used in Lorentz mode electron holography. The hologram width is measured as the distance between the centers of the large Fresnel fringes at each side of the electron hologram. This is different to the field of view that describes the area of the electron hologram captured by the CCD camera. As will be shown, in practice, it is a good idea to match the size of the hologram to the CCD camera.

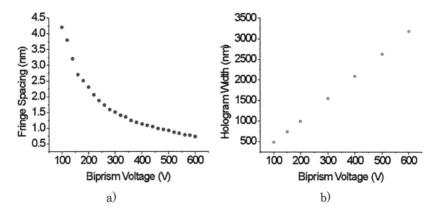

Figure 1.4. *Fringe spacing and field of view of the electron hologram for different biprism voltages typically used in Lorentz mode electron holography*

To perform electron holography, first, the beam alignments must be made as would be performed for high-resolution TEM. The principal difference is that holography is performed using an astigmatic beam. As the holographic fringes are parallel to the biprism wire, coherence is only required in one direction. By deliberately misadjusting the condenser stigmators, highly astigmatic illumination can be obtained that increases the coherent electron flux in the direction of interest. Figure 1.5(a) shows an image of a typical beam setting used for electron holography. Here, one of the stigmators has been set to 100% and the second stigmator has been used to align the major axis of the illumination at exactly 90° to the biprism to maximize the fringe contrast. It is important to take care at this step and make sure that the illumination is well aligned. Figures 1.5(b) and 1.5(c) show the beam focused using C_2 onto the electron biprism at 100% stigmation close to perfect alignment and at perfect alignment, respectively. For perfect alignment, it is clear that the fringes are continuous through the biprism. As the beam is kept astigmatic during the holography experiments, then care must be taken to ensure that the image of the sample is not distorted.

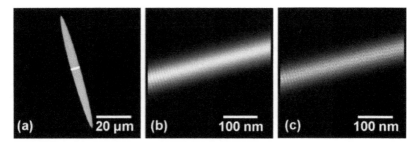

Figure 1.5. *a) An image of the typical stigmation used for electron holography experiments. b) The electron beam has been focused onto the biprism using C_2, both close to perfect alignment and c) at perfect alignment*

The fringe contrast, μ, of the holograms is an important parameter that affects the ultimate phase sensitivity that can be achieved. The fringe contrast is measured using the formula shown below and for Lorentz mode holography; values of 20% or more should be obtained in a carefully aligned electron microscope.

$$\mu = \frac{I_{max} - I_{min}}{I_{max} + I_{min}}$$

1.2.4. Care of the electron biprism

The electron biprism can be easily damaged and it is important to take care of it. Never allow the electron beam to be incident on the biprism unless it is either grounded or connected to the power supply. Also, never focus the beam onto the biprism unless the beam is fully stigmated as this can cause damage. When changing beam settings or magnifications, it is a good idea to remove the biprism from the field of view as the electron beam can be focused down the column by changes in lens currents during these steps. If the biprism is stable and the microscope is performing well for electron holography, then it is a good idea to leave the biprism in the column and instead close the column valve when making changes to the beam settings. It is important to look after the biprism, as a clean and undamaged wire is much more stable and easy to use. For a modern, stable microscope, the biprism is often a weak link, typical biprisms are mechanical and over time the mechanism will become loose. Treat this mechanism with care. When making adjustments to the position of the biprism, if slight mechanical resistance is felt in the mechanism then this suggests that an orientation has been found that can be more stable. Lightly tapping the biprism mechanism will sometimes allow it to stabilize. In addition, the biprism will have some rotational positions that are much more stable than others. It is well worth testing the stability of the biprism with different angles of rotation. Finding stability can be a very frustrating exercise and is certainly not an "exact science", sometimes the microscope will be immediately stable despite changing from different modes and high tensions, and sometimes it can take a few hours to find the best stability.

1.2.5. *Recording electron holograms*

Originally, photographic film was used in electron holography to record the images, combined with an optical reconstruction process. Not only was this approach very time-consuming, but factors such as the nonlinear response of the photographic film and optical artifacts introduced during the reconstruction process introduced additional complications. The development of the CCD camera and powerful computers for hologram reconstruction allows electron holograms to be reconstructed quickly and easily.

The CCD camera has a linear response to incident electrons over a wide dynamic range, which makes it indispensible for quantitative TEM. CCD cameras are typically 1,024–4,096 pixels^2 in size. Cameras with a small number of pixels will limit the number of fringes that can be recorded for a given magnification, thus the field of view. Using a detector with more pixels to improve spatial resolution will be at the expense of intensity and therefore signal to noise. A perfect detector will record every electron in the correct position, but for real detectors, cross-talk between pixels is present. Incident electrons enter the scintillator to generate photons. These photons are then transmitted to the CCD pixels using fiber optics. As the light emission generated in the scintillator can be four times larger than an individual CCD pixel, which is typically 25 µm^2, then a significant amount of cross-talk should be expected. The limitations placed on the spatial resolution from cross-talk should be considered when setting up the microscope. The modulation transfer function (MTF) describes the spatial transfer of information related to the cross-talk of the CCD camera and this can be measured for a given CCD system [DE 95].

The detection quantum efficiency (DQE) measures the efficiency of signal output for each incident electron. A perfect detector will have a DQE of 1. A typical CCD has a DQE of 0.8 at an electron dose of between 1 and 1,000 per pixel, making the CCD very effective at recording incident electrons. Figure 1.6 shows the effect of acquisition time on the formation of an electron hologram. The holograms have been acquired for (a) 0.01 s, (b) 0.1 s, (c) 1.0 s and (d) 10 s. Clearly for an electron microscope using a standard FEG, at least 1 s and preferably more, is required in order to build up an electron hologram on a standard CCD camera. Therefore, dynamic electron holography experiments are at this time limited to time periods of seconds.

Figure 1.6. *Electron holograms with a field of view of 360 nm and a fringe spacing of 2 nm using acquired for time periods of a) 0.01 s, b) 0.1 s, c) 1.0 s and d) 10 s*

The CCD and imaging lenses in the microscope have distortions associated with them. Assuming they are stable, these distortions can be corrected from the hologram by carefully removing the sample from the field of view and obtaining a reference hologram. The complex division of the reconstructed sample and reference waves then reveals a distortion-free phase image. Figure 1.7(a) shows a phase image of an n-channel metal-oxide-semiconductor field effect transistor (nMOS) device specimen reconstructed using only the object hologram. Figure 1.7(b) shows the phase of only in vacuum with the specimen moved out of the field of view. This has been reconstructed from what is known as the reference hologram. In Figure 1.7(c), an nMOS device, reconstructed using both the reference holograms, can be seen and the distortions that are observed in (a) have been removed.

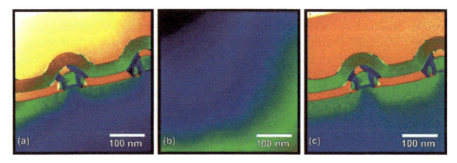

Figure 1.7. *a) Phase image of an nMOS device reconstructed without a reference hologram. b) Phase image reconstructed using an empty reference hologram showing the distortions that are present in the imaging system and CCD camera. c) Phase image reconstructed with the use of the reference hologram*

1.2.6. *Hologram reconstruction*

To recover the information about the amplitude, A, and phase, ϕ, of the electrons, a hologram reconstruction process is required. To reconstruct a hologram with fringe spacing q_c, a Fourier transform is performed. The complex Fourier transform of the hologram is given by the expression:

$$FT\left[I_{hol}(r)\right] = \delta(q) + FT\left[A^2(r)\right]$$
$$+ \delta(q+q_c) \otimes FT\left[A(r)\exp\left[\imath\phi(r)\right]\right]$$
$$+ \delta(q-q_c) \otimes FT\left[A(r)\exp\left[-\imath\phi(r)\right]\right]$$

This expression contains four terms, two peaks at the origin ($q = 0$) corresponding to the Fourier transform of the uniform intensity of the reference image and the Fourier transform of the intensity distribution of the normal TEM image, and two sidebands centered on ($q = -q_c$) and ($q = +q_c$) comprising the Fourier transform of the desired image wavefunction and the Fourier transform of the complex conjugate of the image wavefunction. The two sidebands contain identical information except for a change in the sign of the phase. It is also useful to multiply the hologram by a two-dimensional Hanning window to reduce the appearance of streaks in the Fourier transform arising from the mismatch between the intensity at each edge of the hologram. To obtain the phase image, a sideband is carefully selected and moved to the origin of Fourier space where an inverse Fourier transform is applied. Increasing the size of the mask used to select the sideband will give higher spatial resolution but will also introduce additional noise into the image. As the highest spatial frequencies are not always required, a reduction of the mask radius can be used to remove high-frequency noise. The edges of the mask should be diffuse to minimize any effects due to the abrupt loss of information near its edges. For a strongly scattering object, the maximum radius for the mask is one-third of the carrier frequency, as the radius of the center band is twice that of the sideband. Therefore, the spatial resolution in a reconstructed phase image is typically three times that of the fringe spacing [ALL 99].

If the center of the sideband is not accurately selected, then an artificial gradient can be introduced into the reconstructed phase image. However, the sideband center can be more accurately determined from the Fourier transform of a reference hologram, whose fringe pattern has not been perturbed by the specimen. The same position for the reference sideband and the object sideband can be used. The inverse Fourier transform of the sideband generates a complex image, from which amplitude and phase images can be obtained by using the equations below, where A is the amplitude, ϕ the phase, $\Im$ the imaginary component of the complex reconstructed image and $\Re$ the real component.

$$A(x,y) = \sqrt{\Re^2 + \Im^2}$$

14   Transmission Electron Microscopy in Micro-nanoelectronics

$$\phi(x, y) = \tan^{-1}\left(\frac{\mathfrak{J}}{\mathfrak{R}}\right)$$

Figure 1.8 shows a schematic diagram of the hologram reconstruction process, for a hologram acquired from a symmetrical Si p-n junction containing dopant concentrations of 2×10^{18} cm^{-3}. In the phase image, the n-type region appears brighter than the p-type region. As the dopant concentration is very low compared to the silicon atoms (0.00004%), no contrast is observed in the amplitude image due to the presence of the dopant atoms.

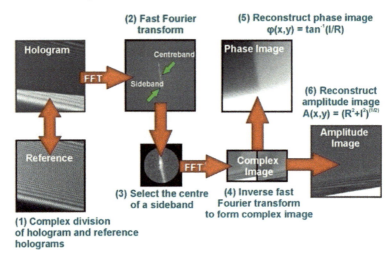

Figure 1.8. *Schematic of the reconstruction procedure used for electron holography*

There are many different software packages that are available for performing the reconstruction of electron holograms, these include Holoworks and HolagraFREE, which are both digital micrograph plug-ins. SEMPER is another software package that contains holography routines and can be used a Linux workstation. All of these packages contain methods to reconstruct and unwrap phase images from an electron hologram. Of course, it is relatively straightforward to perform the appropriate mathematical functions directly to the electron holograms.

One of the more tricky aspects of the reconstruction procedure can be finding the correct sideband to determine the correct sign for the phase images. By convention, the MIP is positive. Therefore, the phase measured in the specimen needs to be positive relative to the vacuum region. This can be complicated for focused ion beam (FIB)-prepared specimens as the rapid change in the thickness from the vacuum to the specimen leads to a phase change of many multiples of 2π in only a few CCD pixels. This makes it difficult to link the measured phase in the sample to the vacuum region. Usually it is possible to find a feature on the sample edge to provide a clue as to which is the correct sideband to use.

1.2.7. Phase Jumps

A phase image is initially reconstructed with phase discontinuities that are unrelated to the specimen features when the phase changes by 2π. Phase jumps occur when the phase change in the region of interest is more than 2π and "phase unwrapping" is required. Figure 1.9(a) shows an electron hologram that has been acquired from a 40° wedge specimen prepared in an FIB containing a p-n junction with a concentration of 2×10^{19} cm^{-3}. A low magnification of 1 k3 has been used to provide a large field of view and the fringe spacing is 4 nm, which leads to a hologram width of 500 nm. A 2,048 × 2,048 pixel CCD camera has been used to sample the fringes correctly and provide a useful holographic field of view of 500 × 1,500 nm. Figure 1.9(b) shows the reference hologram and (c) shows the reconstructed phase image, and the phase jumps of 2π can be clearly observed. In addition, the presence of the p-n junction can be observed in the thicker parts of the specimen. Phase unwrapping has been used to remove the phase jumps in Figure 1.9(d). Profiles extracted from across the indicated regions are shown in Figure 1.9(e) and the phase in multiples of 2π and continuous phase measurement are observed.

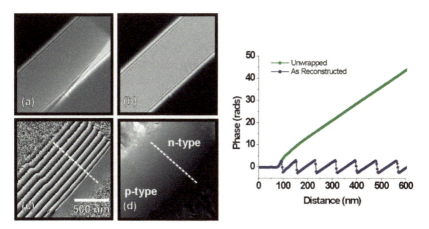

Figure 1.9. *a) Electron hologram of a 40° wedge specimen containing a symmetrical p-n junction with a dopant concentration of 1×10^{19} cm^{-3}. b) Reference hologram, c) reconstructed phase image and d) unwrapped phase image. e) Phase profiles taken from across the region indicated*

1.3. Experimental electron holography

1.3.1. *Fringe contrast, sampling and phase sensitivity*

Experimental electron holography involves achieving a balance between fringe contrast and spatial resolution, hologram width and field of view and the number of electron counts that are recorded. Although the electron source is not perfectly coherent, it must be sufficiently coherent for an interference fringe pattern of sufficient contrast to be acquired within a time period, during which the drift of the specimen, biprism and beam, will not significantly degrade the results.

Considerations regarding the choice of biprism voltage, microscope magnification, electron source size and CCD exposure time will influence the phase information that can be reconstructed from the hologram. The sensitivity of the phase image is associated with the hologram fringe contrast. As has been seen, the ultimate contrast is selected by the choice of beam settings and this can then be maximized by carefully aligning a highly astigmatic electron beam exactly normal to the biprism.

The phase sensitivity is related to the number of electron counts that are recorded and the fringe contrast, μ, by the expression:

$$\Delta\phi_{min} = \frac{\sqrt{2}}{\mu\sqrt{N_{el}}}$$

where N_{el} is the electron dose per pixel in the reconstructed image [LIC 08]. For a given electron beam setting, the fringe spacing can be decreased, and hence the spatial resolution increased by increasing the voltage on the electron biprism. However, by increasing the biprism voltage, two different effects will decrease the contrast. First, the ultimate limit on the achievable spatial resolution is that when higher biprism voltages are used, the virtual sources are pushed further apart and electrons must travel further to form the interference pattern that will reduce the hologram contrast. A second effect that will decrease the contrast is the sampling of the fringes by the CCD camera. When the number of pixels used to sample each fringe starts to fall below six, the recorded hologram contrast will decrease rapidly. Figure 1.10 shows the measured hologram contrast acquired using standard Lorentz lens settings at different biprism voltages to provide fringe spacings in the range 2.5–1 nm for different magnifications and an acquisition time of 16 s. The different magnifications are indicated by the field of view. At higher magnifications, the sampling of the fringes is improved. Using standard beam settings in the Lorentz mode, the smallest fringe spacing is limited to approximately 1 nm before the contrast becomes too low to be useful.

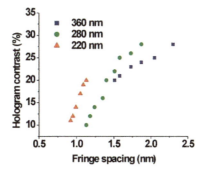

Figure 1.10. *Hologram contrast measured as a function of fringe spacing for different magnifications that are indicated by the total field of view*

Sampling of the electron holograms is a very important aspect of experimental holography and this links the achievable field of view to the spatial resolution. Figure 1.11 shows electron holograms that have been acquired using identical beam settings with a fringe spacing of 2 nm. For the acquisition, a nominal magnification of (a) 2k2 has been used to provide a field of view of 770 nm and (b) 4k7 to provide a field of view of 360 nm. For the higher magnification, each holographic fringe is sampled by six pixels using a $1{,}024 \times 1{,}024$ CCD camera and a contrast of 30% has been measured. For the lower magnification, only 3.4 pixels have been used to sample each fringe and a contrast of only 18% has been measured. Different intensities are observed for the different magnifications as the intensity, C_2, was kept constant during the experiment and using a higher magnification has the effect of spreading the illumination relative to the CCD camera. Whenever possible at least six CCD pixels should be used to sample each holographic fringe, although clearly four pixels can be used at the expense of hologram contrast. Certainly, less than four pixels leads to a rapid decrease in hologram contrast and therefore signal to noise in the reconstructed phase images.

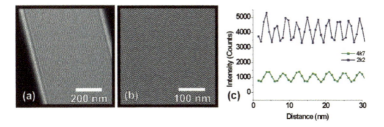

Figure 1.11. *a) Electron hologram with a fringe spacing of 2.0 nm acquired using 3.4 CCD pixels per fringe and b) at higher magnification using six pixels per fringe. c) Profiles acquired from across the fringes showing the reduction in intensity and contrast of the holograms*

To illustrate the difference between beam coherence and fringe sampling, Figure 1.12(a) shows a detail of an electron hologram acquired using a 1,024 pixel acquisition to provide a field of view of 360 nm. The fringe spacing of 2.2 nm is correctly sampled by 5.7 pixels per fringe, which allows a contrast of 30% to be preserved. Figure 1.12(b) shows the same acquisition settings except for a 1.6-nm fringe spacing. Now the fringe sampling is only 4.2 pixels and the contrast has been reduced to 15%, due to both the coherence of the beam and the sampling of the fringes. In Figure 1.12(c), the same hologram can be seen, except now acquired using a 2,048 pixel acquisition, the sampling has increased to 8.4 pixels per fringe and a contrast of 22% is now recorded. Finally, Figure 1.12(d) shows that when further increasing the biprism voltage to provide a 1.2-nm fringe spacing, although the fringes are still correctly sampled at 6.5 pixels per fringe, the contrast has dropped to 15% due to the limited coherence of the electron beam.

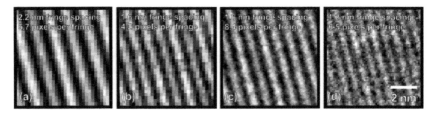

Figure 1.12. *Detail of the center of an electron hologram with identical magnifications to provide a total field of view of 360 nm but with different fringe spacings and sampling densities*

The sampling of the electron hologram fringes has important implications for electron holography when low magnifications are required. The hologram width and fringe spacing are linked and a fringe spacing of 4 nm is about the largest that can be obtained while still providing a hologram that is wide enough to be useful. If conventional Lorentz beam settings and a 1,024 × 1,024 CCD camera are used and four pixels are required to sample each fringe, then it is clear that one pixel is required per nanometer leading to a maximum field of view that can be achieved in Lorentz mode of approximately 1.0 μ. This can be increased by using a 2,048 × 2,048 pixel CCD camera at the expense of electron counts.

To illustrate the sensitivity that can be expected by using different acquisition times and typical beam settings, two different empty holograms were recorded using identical settings for different times and used as the object and reference holograms. Figure 1.13(a)–(d) shows electron holograms that have been acquired in the Lorentz mode with a fringe spacing of 2 nm and a field of view of 360 nm using the acquisition times indicated. The reconstructed phase images are shown in Figure 1.13(e)–(h). To give an idea of the contrast, profiles extracted from across the electron holograms are shown in Figure 1.13(i). In order to illustrate the noise in the phase images, profiles taken from across the center of the images are shown in Figure 1.13(j). As seen previously, for a perfectly stable electron microscope, the signal to noise is dependent on the hologram contrast and the number of electron counts that are recorded. In Figure 1.13(k), the experimentally measured sensitivity of the phase images is shown. The sensitivity has been assessed by taking the standard deviation of the center region of each of the phase images. The stability of the electron microscope has allowed the holograms to be acquired for time periods of between 1 and 64 s with a fringe contrast of approximately 40% irrespective of the acquisition times. These have been compared to the sensitivity calculated by theory and although the fit is not perfect, the link between the hologram contrast, the number of counts that are recorded in the hologram and the size of the mask that is used to reconstruct the hologram (required spatial resolution) is clear. As a result, it is possible to estimate the beam settings that will be required in order to provide the sensitivity to successfully perform a desired experiment.

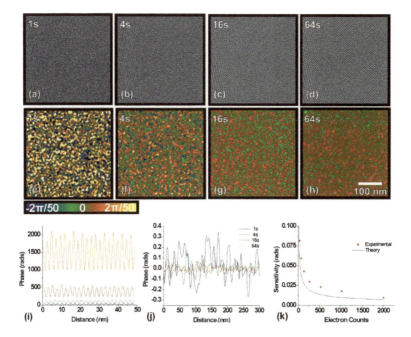

Figure 1.13. *(a)–(d) Electron holograms that have been acquired for time periods in the range 1–64 s. (e)–(h) Phase images reconstructed from two electron holograms containing only the vacuums that have been acquired directly after one another, the first as an object hologram, the second as a reference. i) Profiles of electron holograms shown in (a)–(d) revealing the contrast. j) Profiles of phase images reconstructed using two empty holograms containing only vacuum shown in (e)–(h). k) The experimentally determined and calculated sensitivities plotted as a function of the average number of electron counts recorded on the CCD camera*

The phase sensitivity measured experimentally is $2\pi/76$, $2\pi/150$, $2\pi/280$ and $2\pi/600$ for the holograms acquired for 1, 4, 16, and 64 s, respectively. Values of phase that can be expected when performing electron holography at 200 kV for different types of samples are as follows. In a perfect and undamaged 200-nm-thick specimen containing a p-n junction in a modern device structure, we would expect to see a step in potential of about 1 V, which corresponds to a step in phase of approximately $2\pi/4$ [RAU 99]. The phase change caused by single atoms approximately follows the relationship $\delta\phi = Z^{0.6}$, and for a single gold or silicon atom we would expect to measure a step in phase of around $2\pi/10$ and $2\pi/50$, respectively [LIC 08]. To measure the step in phase of a single ionized dopant atom would be significantly more difficult. Assuming that a screened potential of a phosphorus atom in silicon can have a potential of a few hundred millivolts across a radius of 1 or 2 nm, a phase change of $2\pi/1,000$ could be expected [KOH 55].

Using a modern TEM in the medium resolution Lorentz mode, we can expect to obtain a sensitivity of $2\pi/1,000$ relatively easily [LIN 12]. However, in high-resolution mode, a best sensitivity of around $2\pi/250$ has been recently demonstrated [COO 07]. Certainly, there is more work to be done for both instrumentation and specimen preparation before individual ionized dopant atoms can be detected with atomic scale resolution.

1.3.2. *Optimizing the beam settings for an electron holography experiment*

When planning an electron holography experiment, it is important to assess the spatial resolution and field of view that is required, as well as the minimum value of signal to noise needed in order to measure the changes in electrostatic potential. Using standard lens settings in the microscope where only the magnification and focus are adjusted, the spatial resolution and the field of view are fixed by the biprism voltage. The choice of beam settings (coherence) will also determine the achievable spatial resolution. Usually, in order to obtain a very good spatial resolution, a weak electron beam is required where the electrons that are collected are only emitted from the very center of the electron gun. This can be controlled by using combinations of a low extraction voltage, a low gun lens setting and a large spot size and a small C2 aperture. There is no "correct" setting for the holography mode, the author tends to prefer to use a very weak electron beam combined with longer acquisition times. For FEI microscopes, an extraction voltage of 3,810 V, a gun lens setting of 5, a spot size of 3 and a 150 µ C2 aperture are good settings to begin with, if more counts are required then the spot size can be reduced, which will double the number of electron counts available per step. Modern microscopes equipped with a monochromator are excellent for electron holography, as the intensity of the electron beam can be flexibly adjusted using the monochromator focus if more counts are needed. It is important to be aware that intense electron beams do not only lead to a reduction in the hologram contrast, but can also cause the buildup of charge in the specimens, which can affect the phases that are measured in doped semiconductor specimens.

Most modern electron microscopes can be operated in the range 300–80 kV and sometimes even lower. The choice of operating high tension is generally dependent on the stability of the specimen under the electron beam. From Figure 1.1 it can be seen that the electrons interact more strongly with the specimen at low energies. For the examination of Si semiconductor specimens, an operating voltage of 200 kV provides a good compromise between reducing specimen damage, providing a large phase change in the electrons and also providing enough electron counts to be transmitted through the specimen. Electron holography can also be performed at lower high tensions; however, as thinner specimens can be required, this has implications for the dopant concentrations that can be measured. In addition, the gun can be operated using different settings to provide brighter electron beams or more coherent beams. Figure 1.14 shows the contrast measured across electron holograms acquired with a fringe spacing of 3 nm and a field of view of 750 nm as a function of the extraction voltage and operating high tension. Here, the gun lens and spot size have been kept constant. There is a large difference between the number of recorded counts and contrast for the different settings.

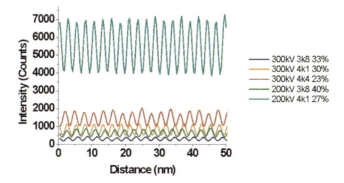

Figure 1.14. *Profiles extracted from across different electron holograms for different beam settings and high tensions. All of the holograms were acquired for 16 s using a nominal magnification 2k9 to provide a field of view of 750 nm with a fringe spacing of 3 nm. The measured contrast is also indicated*

1.3.3. *Optimizing the field of view using free lens control*

A different approach to improve the maximum spatial resolution that is achievable is to use free lens control. Although the biprism voltage can be increased to improve the spatial resolution, the hologram contrast will deteriorate and for many holography electron microscopes, it is difficult to have fringe spacings much below 2 nm, which limits the spatial resolution to approximately 5 nm. Figure 1.15(a) shows an image of a standard beam configuration with the electron beam fully stigmated and 200 V applied to the biprism. The associated hologram in (b) has been acquired using a nominal magnification of 4k7 to provide a field of view of 360 nm and the each fringe is sampled by six pixels. If a better spatial resolution is required, then a higher voltage can be applied to the biprism. Figure 1.15(c) shows an image of the beam with 400 V applied to the biprism to provide a fringe spacing of 1 nm. However, now that the interference pattern is much wider, therefore even if the hologram is correctly sampled such as in (d), which shows an electron hologram that has been acquired using a nominal magnification of 8k0 and a field of view of 200 nm, the fringes will have lost contrast due to the limited coherence of the beam. Figure 1.15(e) shows an image of the beam configuration after selecting a high magnification in the Lorentz mode and then using the diffraction lens to reduce the width of the interference pattern. Here, there is still 400 V applied to the biprism; however, as the width of the pattern has been reduced, the contrast of the holograms is still high. Figure 1.15(f) shows an electron hologram acquired using this configuration with a field of view of 150 nm and a fringe spacing of 1 nm. Figure 1.15(g) shows profiles extracted from the holograms in Figures 1.15(b), (d) and (f). By using free lens control, it can be seen that 1-nm fringe spacing can be achieved with a good contrast level and with enough electron counts. When higher magnifications and spatial resolutions are required such that the electron hologram is larger than the field of view, the diffraction lens should always be used to reduced the width of the electron hologram so that the Fresnel fringes are just outside the field of view of the CCD camera in order to maximize the fringe contrast and hence the sensitivity.

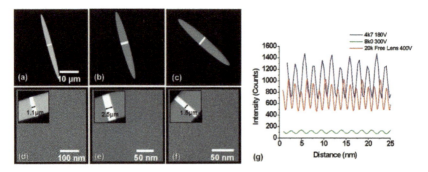

Figure 1.15. *a) An image of the electron beam used to acquire the electron hologram shown in b), here a standard Lorentz magnification is used with a biprism voltage of 200 V to provide a fringe spacing of 2.0 nm and an overlap width of 1.1 µ. c) An image of the electron beam used to acquire the hologram shown in d) with 400 V applied to the biprism to achieve a fringe spacing of 1.2 nm and an overlap with of 2.5 µ. e) By adjusting the diffraction lens settings, the interference width can be reduced while preserving the fringe spacing. f) An electron hologram with a fringe spacing of 1 nm and a total width of 1.5 µ. g) Profiles acquired from all of the holograms showing intensity and the fringe contrast*

There are many different methods that can be used to improve the spatial resolution when performing electron holography experiments. On both JEOL and FEI microscopes, a dual lens method can be used where the Lorentz and objective lenses are used simultaneously. This approach provides fringe spacings of approximately 1 nm [WAN 04]. An important consideration is that for high spatial resolutions, the field of view is limited, which means that a vacuum region must be provided very close to the region of interest. For very small samples, conventional high-resolution electron holography can be used to provide a field of view of a few tens of nanometers and atomic resolution.

An example of the benefits of using the diffraction lens is shown. Figure 1.16(a) shows an STEM image of an nMOS device with a 28-nm gate length in (b). The corresponding potential map that has been acquired using conventional beam settings to provide a field of view of 360 nm and a fringe spacing of 2 nm can be seen. However, it is difficult to determine the position of the active dopant atoms and the electrical junction under the gate. The potential map shown in Figure 1.16(c) has been acquired by adjusting the diffraction lens to provide a field of view of 150 nm and a fringe spacing of 1 nm. Now details such as the SiGe channel and the electrostatic potentials under the channel arising from the doped source and drain regions can be clearly seen. To achieve this spatial resolution, a vacuum reference region is needed near the region of interest, here the specimen was rotated by 90° to the ion beam in the FIB after lift out and the metal surface layers were removed to provide a vacuum reference 150 nm from the region of interest. If higher magnifications are required then fields of view in the range of 30–100 nm are also possible, by applying free lens control in objective lens mode. Objective mode holography is usually associated with atomic resolution imaging that provides fields of view below 30 nm. Therefore, the goal is the opposite as for the Lorentz mode and a wider interference pattern and larger fringe spacings are needed. By adjusting the diffraction lens and simultaneously increasing the projector lenses, the required hologram properties can be obtained [SIC 11]. In Figure 1.16(d), a potential map of an arsenic-doped pMOS device with a 40-nm gate length can be seen. To be able to accurately measure the electrical gate length, better spatial resolution is required as the region of interest is small compared to the total field of view. Figure 1.16(e) shows a potential map that has been acquired using the free lens objective mode and a field of view of 75 nm, and a fringe spacing of 0.33 nm has been achieved to provide a spatial resolution of 1 nm. The electrical overlap width, dl, has been compared to the arsenic distribution measured using electron energy loss spectroscopy (EELS) and the results are consistent within the spatial resolutions of the two techniques. However, to achieve these results, it was not possible to put a vacuum reference region close enough to the region of interest without damaging the specimen in the FIB. As a result, the lightly doped substrate has been used as a reference. Using a homogeneous region of the specimen instead of the vacuum reference is known as "differential" or "bright field" electron holography.

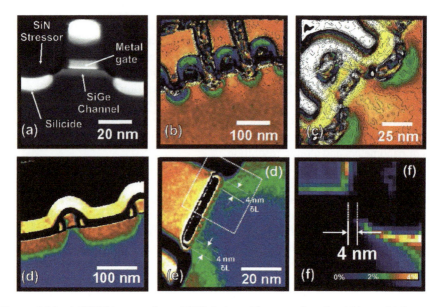

Figure 1.16. *a) STEM image of an nMOS device with a gate length of 28 nm. b) Potential map showing the doped regions with a spatial resolution of 6 nm and a field of view of 360 nm. c) Potential map with a spatial resolution of 3 nm and a field of view of 150 nm provided in a free lens Lorentz mode. d) Potential map of a pMOS device with a gate length of 40 nm. The spatial resolution is 6 nm and the field of view 360 nm. e) Potential map of the pMOS device with a spatial resolution of 1 nm and a field of view of 75 nm. f) EELS map of the region indicated in e) showing the concentration of arsenic atoms*

1.3.4. *Energy filtering for electron holography*

An energy filter can also be used to improve the contrast in the object hologram as this will remove the diffuse background scattering in the recorded image. Figure 1.17(a) shows an electron hologram containing an nMOS device acquired with a 10 eV slit inserted and (b) the potential map calculated from the reconstructed phase image. In Figure 1.17(c), an unfiltered electron hologram and (d) the potential map are also shown. Profiles acquired from the indicated regions in the electron holograms are shown in Figure 1.17(e). The hologram contrast in the silicon region of interest is improved from a value of 4% in the unfiltered hologram to 6% in the filtered hologram. For a comparison between the fringe contrasts between the region of interest and the vacuum region, a profile of the reference hologram is also shown. The improvement can clearly be seen in the phase images. The standard deviation has been measured in the regions indicated in Figures 1.17(b) and (d), and 0.10 rads is measured in the unfiltered phase image and 0.08 rads is measured in the filtered phase image. An energy filter will provide more significant improvements for thicker TEM specimens and is also extremely valuable when using dark-field electron holography for strain mapping.

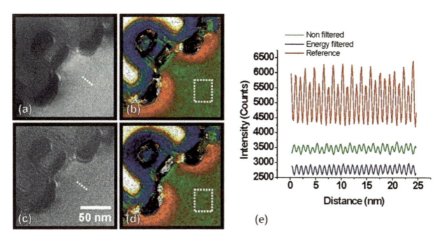

Figure 1.17. *a) Filtered electron hologram containing a 210-nm-thick nMOS device. b) Electrical potential map calculated from (a). c) Unfiltered electron hologram and d) electrical potential map. e) Profiles showing the fringe contrast acquired from the indicated regions in (a) and (b) and from the reference hologram*

1.3.5. *Minimizing diffraction contrast*

The measured phases in electron holography are extremely sensitive to the geometrical crystal potential, ϕ^G and as a result it is important to orient the specimen to a weakly diffracting condition. This has important consequences for electron holography as tilting the specimen from the zone axis can lead to a loss in the spatial resolution in the projection of the specimen. The specimen can be examined at an orientation of choice, which means that the spatial resolution imposed by the tilt can be eliminated for either the alpha or beta tilt direction. When performing electron holography on a specimen, it is important to find the zone axis and then tilt the specimen along one of the zone axes by 1° or 2°, or less if possible. Fine specimen tilts can then be applied to remove all traces of diffraction in the specimen and the contrast in the region of interest should be completely homogeneous and bright. In Figure 1.18(a), an electron hologram containing an nMOS device in a diffracting orientation can be seen. The diffraction contrast is also present in the reconstructed phase and amplitude images shown in Figures 1.18(b) and (c), which will lead to misleading measurements of the specimen potential. In Figure 1.18(d), an electron hologram of a different device, but from the same lamella tilted to a more weakly diffracting orientation, is shown. The diffraction contrast has been significantly reduced from the hologram, and the phase and amplitude images in (e) and (f) have much more homogeneous contrast. It is not always possible to remove all the diffraction contrast from specimens during examination; however, this should be minimized and the effects of diffraction on the measured phase can be assessed by acquiring holograms with different amounts of contrast present.

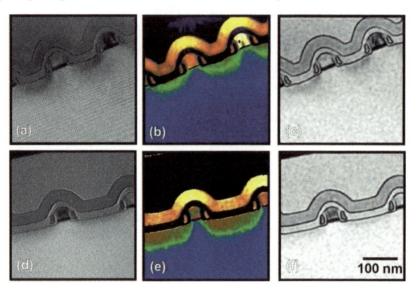

Figure 1.18. *a) Electron hologram containing an nMOS device with diffraction contrast in the source and drain regions. b) and c) Reconstructed phase and amplitude images showing evidence of diffraction contrast. d) Electron hologram containing an nMOS device tilted to a weakly diffracting orientation. e) and f) Reconstructed phase and amplitude images showing more homogeneous contrast in the region of interest*

1.3.6. *Measurement of the specimen thickness*

The phase change of an electron is determined by the integral of the potentials in and around the specimen. Therefore, in order to recover information about the potentials, it is important to know the thickness of the specimen. There are many different methods that can be used to measure specimen thickness; here, we focus on two that are most commonly used in electron holography. Convergent beam electron diffraction (CBED) can be used to determine the crystalline thickness of a specimen to approximately ±5 nm [WIL 09]. Figure 1.19(a) shows a CBED pattern that has been acquired for a thick specimens (>150 nm), here a CBED pattern in a two beam condition will provide two discs from which the specimen thickness can be determined from the distances between the interference fringes. For thinner specimens, a single CBED pattern can be acquired at the zone axis and compared to simulations to determine the correct thickness. If CBED is performed carefully, and the patterns are well focused, then this provides a robust and reproducible measure of the specimen thickness.

An alternative method of measuring the thickness is to use the reconstructed amplitude image. From electron energy loss spectroscopy, it is known that the image intensity is related to the inelastic mean free path, λ_{in} [EGE 11].

$$I_0 = I_{total} e^{-t/\lambda_{in}}$$

As the amplitude image is the square root of the intensity, an expression can be derived that relates the specimen thickness to the amplitude image [MCC 94].

$$\frac{t(x,y)}{\lambda_{in}} = -2\ln A(x,y)$$

Therefore, if the sample is homogeneous and the effects of dynamical diffraction are limited, then the thickness of the specimen can be calculated in units of λ_{in}. Figures 1.19(b), (c) and (d) show an electron hologram, amplitude image and thickness map, respectively, that has been calculated for a silicon lamella of thickness 400 nm. When calculating the thickness from the amplitude image, the image must first be normalized with respect to the vacuum region. Also, λ_{in} is dependent on microscope parameters such as the objective aperture size. To determine λ_{in} for Si using the FEI Titan, the thickness of five different lamellas prepared by FIB milling at 30 kV was evaluated by CBED and from the amplitude images. Figure 1.19(e) shows the relationship between the crystalline thickness of the specimen and its total thickness in units of λ_{in}. From the gradient, λ_{in} is calculated as 120 ± 10 nm for these microscope settings. From the x-intercept, the total thickness of the amorphous layer can be determined as 50 nm (25 nm on each face), which is consistent with what would be expected for an FIB operating voltage of 30 kV. Different values of λ_{in} have been given elsewhere that have been determined using different microscopes, operated at different high tensions using specimens prepared by different methods. Clearly, care needs to be taken when using this method of calculating the thickness of specimens. This is especially valid when thicker specimens are being examined where multiple scattering of the electrons is likely, which is often the case for off-axis electron holography.

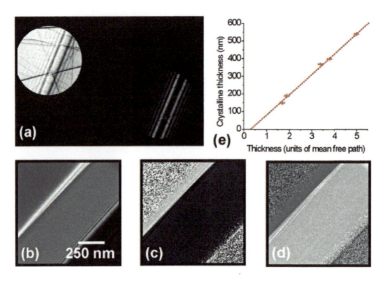

Figure 1.19. *a) CBED pattern of a 400-nm-thick specimen in a (004) two-beam condition. b)– d) An electron hologram, amplitude image and thickness map, respectively, of a 400-nm-thick silicon specimen. e) The crystalline thickness of a range of silicon specimens plotted against the thickness determined from the amplitude image*

1.3.7. *Specimen preparation*

Specimen preparation is the key to electron holography. The stability and computer control of modern TEMs make experimental electron holography relatively straightforward to perform. However, as has been discussed, changes in the MIP are typically in an order of magnitude higher than the electrical potential in a junction. As a result, excellent, flat specimens are required otherwise the data becomes much more difficult to interpret.

The process of using either Ar or Ga ions to thin a sample for observation by TEM will introduce many artifacts into the specimen that can change the measured potentials. Ideally, all specimens would be made by polishing as this provides specimens that have relatively undamaged surfaces. Examples of electron holography performed on wedge polished semiconductor devices do exist in the literature [TWI 02]; however, it is difficult and time consuming to find a specific semiconductor device with a gate length of 40 nm or less in a 300-mm wafer. In addition, a vacuum region is required to be close to the region of interest adding a further complication. In practice, most specimens that are made by polishing are also Ar milled to clean the region of interest, and even the use of low-energy ions can introduce significant artifacts when performing dopant profiling.

FIB milling allows parallel-sided specimens to be prepared from a region of interest with little difficulty. Modern dual-beam FIBs make it relatively easy to find an individual semiconductor device by using the scanning electron microscope. FIBs use a focused beam of Ga ions to cut around the region of interest. The region of interest is then extracted by *in situ* lift-out and then thinned to electron transparency. The strengths of the FIB are that it allows the thickness to be controlled and also allows parallel-sided specimens to be prepared, which do not have thickness variations across the region of interest. The problems with amorphization on specimen surfaces are well known; however, for electron holography this is not a great problem. In fact, the presence of the surface amorphous layers will create an isopotential surface on the specimen that will eliminate fringing fields in the vacuum regions near the specimen. The real problem of using Ga ion milling is that the ions penetrate the specimen and create what is known as the inactive thickness, t_{inactive}, which will be discussed later in this chapter.

Semiconductor specimens usually contain insulators between the silicon substrate and the metalized regions on the surface. Figure 1.20(a) shows a phase image of a pMOS device with the insulator region charging during observation in the TEM. The charging has completely dominated the dopant potentials in the specimen. Figure 1.20(b) shows a phase image of an nMOS device that has been carbon coated, but not prepared using backside milling. The device has no metallization on the top layers and the vertical stripes in the phase image are caused only by changes in the composition between the silicon, silicon oxide and silicon nitride. For a metallized device, the problems caused by differential milling would be significantly worse. Figure 1.20(c) shows the same device as seen in 1.20(a). The specimen has been prepared using backside milling and a thin layer of carbon has been applied onto one side of the specimen. The effects of charging have been significantly reduced and the changes in the potentials arising from the active dopant atoms can be clearly seen.

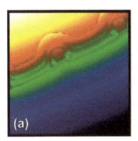 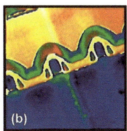 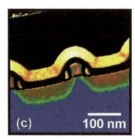

Figure 1.20. *a) Phase image of nMOS device without carbon coating. b) Phase image of a carbon-coated nMOS device that has been prepared using conventional FIB milling. c) Phase image of an nMOS device that has been prepared by back side FIB milling and then carbon coated*

1.3.8. *The electrically inactive thickness*

The relationship between the measured phase and the potential inside the semiconductor specimens is complicated due to the effects of specimen preparation and surface charging, which lead to what is known as the inactive thickness. There are many papers that discuss the problems encountered when using electron holography for dopant profiling; in this chapter, there is only space to introduce them briefly. The inactive thickness is used to describe a modified crystalline region in the specimen where the dopant atoms are not active [GRI 02]. From the equation below, it can be seen that in principle, it should be straightforward to recover the electrical potential across an electrical junction, where:

$$V_E = \frac{\Delta \phi}{C_E \cdot t_{active}}$$

however, the problem is the determination of the specimen thickness that contains active dopants, t_{active}. Clearly, the amorphous surface layers can be neglected from the calculation of the potential as these will not contain active dopants. Figure 1.21 shows a phase image of a Si specimen containing a p-n junction with a symmetrical dopant concentration of 2×10^{18} cm^{-3} that has been prepared by FIB milling at 30 kV. It is clear that the electrical junction does not extend all the way to the specimen surface, indeed the junction becomes visible at about 100 nm from the specimen edge. If a series of these p-n junctions with different thicknesses are examined and the step in phase across the electrical junction is measured such as in Figure 1.21(b) and then evaluated as a function of the crystalline thickness measured by CBED as in Figure 1.21(c), then the inactive thickness can be determined from the x-intercept. Here, the value of the inactive thickness is 140 nm. The inactive thickness will be equally distributed on each specimen surface; thus, 70 nm inactive thickness on each surface plus an amorphous layer of approximately 25 nm that corresponds well with the phase image in (a). Figure 1.21(d) shows a schematic of a TEM specimen containing a p-n junction.

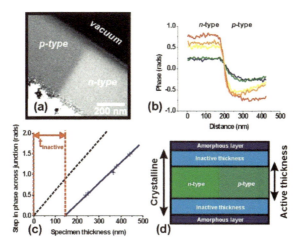

Figure 1.21. *a) Phase image of a specimen containing a p-n junction with a concentration of 2×10^{18} cm^{-3}. b) Phase profiles extracted from across different specimens of a range of thicknesses. c) The step in phase measured across the p-n junctions as a function of the crystalline specimen thickness measured by CBED. d) Schematic of a TEM specimen containing a p-n junction*

If the inactive layer is known then the potential in a p-n junction can be calculated. The problem is that the inactive thickness is sensitive to the dopant concentration in the specimens, the amount of charge and also specimen preparation [COO 10]. Therefore, it is very difficult to predict the inactive thickness. Figures 1.22(a)–(c) show the experimentally measured phase across 400-nm-thick specimens containing p-n junctions with a range of dopant concentrations. For low dopant concentrations, there is a larger discrepancy between the experimentally measured values of phase and theory. This is demonstrated more clearly in Figure 1.22(d) where the step in phase across a series of junctions with different dopant concentrations is shown as a function of the crystalline specimen thickness. Here from the *x*-intercept it is clear that the inactive thickness varies strongly with the dopant concentration. The inactive thickness has been measured from the intercepts in Figure 1.22(d) and is shown as a function of the dopant concentration in Figure 1.22(e). Here, we see that the inactive thickness tends toward zero for higher dopant concentrations and can be reduced by using lower FIB operating voltages. The inactive thickness is caused by a combination of band bending caused by the presence of charge at the specimen surfaces and also from the effects of specimen preparation. The use of ions to prepare specimens for TEM analysis will create defects deep in the specimens and trap the dopants. These effects can be reduced by reducing the energy of the ions or by using larger ions that do not penetrate deep in the specimen. For example, Figure 1.22(e) also shows the inactive thickness measured in specimens prepared using an FIB operating voltage of 8 kV instead of 30 kV, and the inactive thickness has been reduced by a factor of two.

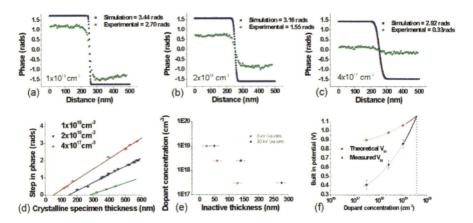

Figure 1.22. a), b) and c) The experimentally determined step in phase compared with simulations for specimens containing p-n junctions with different dopant concentrations. d) The step in phase across a range of p-n junctions as a function of the crystalline specimen thickness. e) The inactive thickness measured experimentally for specimens with different dopant concentrations that have been prepared using different FIB operating voltages. f) The measured built in potential across different p-n junctions compared to theory

Finally, the electrical potentials can be determined from the gradients in Figure 1.22(d), which uses a method that is in principal independent of the inactive thickness, where,

$$\phi = C_E V_E (t_{cryst} - t_{inactive})$$

The calculated values of the built in potential in the different p-n junctions are shown in Figure 1.22(f) and are compared to the theoretically expected values. Again, the experimental results suggest that for higher dopant concentrations the experimentally determined values tend toward theory. Again, suggesting that by examining samples with high dopant concentrations, the artifacts that are observed become less important.

If the detection of very low dopant concentrations is required, then thick specimens that are prepared extremely carefully using low-energy ions, or only polishing, are needed. Low dopant concentrations are very sensitive to surface charging and the whole thickness of the specimen can be depleted [COO 08]. In fact, the observation of dopant concentrations below 1×10^{17} cm^{-3} in silicon is very difficult. Because of the reduction in the size of modern semiconductor devices and the very high dopant concentrations that are present, it is becoming more straightforward to be able to determine information about the position of the electrical junctions in the devices.

The literature is full of approaches such as in situ annealing of Si specimens at low temperature (300°C) or the use of large ions such as Xe at low energy in order to significantly reduce the inactive thickness. In situ biasing can also be used to recover the correct built in potential. Although it is still difficult to obtain quantitative information about the dopant concentrations in semiconductor devices, approaches such as these may allow fully quantitative dopant profiling in the near future.

1.4. Conclusion

Off-axis electron holography provides the opportunity to measure the phase change of an electron wave to provide maps of the magnetic, strain and dopant potentials with nanometer-scale resolution. The stability of modern electron microscopes now makes off-axis electron holography a relatively straightforward technique to perform. The electron beam can be quickly and easily aligned and the biprism can be inserted to provide excellent contrast electron holograms that can be recorded with many electron counts. The key for electron holography is to have excellent quality, flat specimens. This is especially true for dopant profiling, as the MIP of most materials is an order of magnitude larger than the potentials expected from active dopants. Therefore, if the specimen is not flat, then it will be extremely difficult to obtain any useful information about the dopants in the specimens. The success of failure of an experiment will be determined before the specimen is put into the microscope. However, we hope that the examples of the dopant maps of real transistors that are shown here are enough to convince the reader of the possibilities of off-axis electron holography. It has not been possible to review all of the different examples of electron holography in the space available here. Groups based in Dresden, Arizona, Bologna and Japan have led the way, and now the technique is becoming widespread and centers can be found in Berlin, Jülich, Toulouse and Grenoble. The literature is now full of wonderful examples of electron holography being used to assess the electrical potentials in transistor devices, nanowires and quantum dots. Certainly, the prospects for off-axis electron holography for strain mapping, and for the visualization of magnetic fields and dopant potentials make it unique and it will surely become an indispensible tool for the characterization of nano-scaled materials.

1.5. Bibliography

[ALL 99] ALLARD L.F., VOLKL E., JOY D., *Introduction to Electron Holography*, Plenum, New York, NY, 1999.

[BET 28] BETHE H., "Theorie der Beugung von Elektronen au Kristallen", *Annalen der Physik*, vol. 87, pp. 55–129, 1928.

[COO 07] COOPER D., TRUCHE R., RIVALLIN P., HARTMANN J-M., LAUGIER F., BERTIN F., CHABLI A., ROUVIÈRE J-L., "Medium resolution off-axis electron holography with millivolt sensitivity", *Applied Physics Letters*, vol. 91, 143501, 2007.

[COO 08] COOPER D., AILLIOT C., TRUCHE R., BARNES J-P., HARTMANN J-M., BERTIN F., "Experimental off-axis electron holography of focused ion beam prepared Si p-n junctions with different dopant concentrations", *Journal of Applied Physics*, vol. 104, 064513, 2008.

[COO 10] COOPER D., AILLIOT C., BARNES J-P., HARTMANN J-M., SALLES P., BENASSAYAG G., DUNIN-BORKOWSKI R.E., "Dopant profiling of focused ion beam milled semiconductors using off-axis electron holography", *Ultramicroscopy*, vol. 110, pp. 383–389, 2010.

[COO 11] COOPER D., DE LA PENA F., BÉCHÉ A., ROUVIÈRE J-L., SERVANTON G., PANTEL R., MORIN P., "Field mapping with nanometer scale resolution for the next generation of electronic devices", *Nano Letters*, vol. 11, pp. 4585–4590, 2011.

[DER 95] DE RUIJTER W.J., "Imaging properties and applications of slow-scan charge coupled device cameras suitable for electron microscopy", *Micron*, vol. 26, pp. 247–275, 1995.

[EGE 11] EGERTON R.F., *Electron Energy Loss Spectroscopy in the Electron Microscope*, Springer, Berlin, 2011.

[GAB 48] GABOR D., "A new microscopic principle", *Nature*, vol. 161, pp. 777–778, 1948.

[GRI 02] GRIBELYUK M., MCCARTNEY M.R., LI J., MURTHY C.S., RONSHEIM P., DORIS B., MCMURRAY J.S., HEGDE S., SMITH D.J., "Mapping of electrostatic potential in deep submicron CMOS devices by electron holography", *Physical Review Letters*, vol. 89, 025502, 2002.

[HYT 11] HYTCH M., HOUDELLIER F., HUE F., SNOECK E., "Dark field electron holography for the measurement of geometric phase", *Ultramicroscopy*, vol. 111, pp. 1328–1337, 2011.

[KOH 55] KOHN W., LUTTINGER J.M., "Theory of donor states in silicon", *Physical Review*, vol. 98, pp. 915–922, 1955.

[KRU 06] KRUSE P., SCHOWALTER M., LAMOEN D., ROSENAUER A., GERTHSON D., "Determination of the mean inner potential in III-V semiconductors, Si and Ge by density functional theory and electron holography", *Ultramicroscopy*, vol. 106, pp. 105–113, 2006.

[LIC 07] LICHTE H., FORMANEK P., LENK A., LINCK M., MATZECK C., LEHMANN M., SIMON P., "Electron holography, applications to materials questions", *Annual Review of Materials Research*, vol. 37, pp. 539–588, 2007.

[LIC 08] LICHTE H., "Performance limits of electron holography", *Ultramicroscopy*, vol. 108, pp. 256–262, 2008.

[LIN 12] LINCK M., FREITAG B., KUJAWA S., LEHMANN M., NIERMANN T., "State of the art in atomic resolution off-axis electron holography", *Ultramicroscopy*, vol. 116, pp. 13–23, 2012.

[MCC 94] MCCARTNEY M.R., GAJDARDZISKA-JOSIFOVSKA M., "Absolute measurement of normalised thickness from off-axis electron holography", *Ultramicroscopy*, vol. 53, pp. 283–289, 1994.

[MCC 07] MCCARTNEY M., SMITH D., "Electron holography: phase imaging with nanometer resolution", *Annual Review of Materials Research*, vol. 37, pp. 729–767, 2007.

[MIS 81] MISSIROLI G.F., POZZI G., VALDRÈ U., "Electron interferometry and interference electron microscopy", *Journal of Physics E Scientific Instruments*, vol. 14, pp. 649–671, 1981.

[ORC 95] ORCHOWSKI A., RAU W.D., LICHTE H., "Electron holography surmounts resolution limit of electron microscopy", *Physical Review Letters*, vol. 74, pp. 399–402, 1995.

[RAD 70] RADI G., "Crystal lattice potnetials in electron diffraction calculated for a number of crystals", *Acta Crystallographica*, vol. A26, pp. 41–56, 1970.

[RAU 99] RAU W.D., SCHWANDER P., BAUMANN F.H., HOPPNER W., OURMAZD A., "Two-dimensional mapping of the electrostatic potential in transistors by electron holography", *Physical Review Letters*, vol. 82, pp. 2614–2617, 1999.

[REI 89] REIMER L., *Transmission Electron Microscopy*, Springer, Berlin, 1989.

[REZ 94] REZ D., REZ P., GRANT I., "Dirac-Fock calculations of X-ray scattering factors and contributions to the mean inner potential for electron scattering", *Acta Crystallographica*, vol. A50, pp. 481–497, 1994.

[SIC 11] SICKMANN J., FORMANEK P., LINCK M., MUEHLE U., LICHTE H., "Imaging modes for potential mapping in semiconductor devices by electron holography with improved lateral resolution", *Ultramicroscopy*, vol. 111, pp. 290–302, 2011.

[TON 87] TONOMURA A., "Applications of electron holography", *Reviews of Modern Physics*, vol. 59, pp. 639–669, 1987.

[TWI 02] TWITCHETT A.C., DUNIN-BORKOWSKI R.E., MIDGLEY P.A., "Quantitative electron holography of biased semiconductor devices", *Physical Review Letters*, vol. 88, 238302, 2002.

[WAN 04] WANG Y.Y., KAWASAKI M., BRULEY J., GRIBELYUK M., DOMENICUCCI A., GAUDIELLO J., "Off-axis electron holography with a dual lens imaging system and its usefulness in 2-D potential mapping of semiconductor devices", *Ultramicroscopy*, vol. 101, pp. 63–72, 2004.

[WIL 09] WILLIAMS D.B., CARTER B., *The Transmission Electron Microscope*, Springer, Berlin, 2009.

Chapter 2

Dopant Distribution Quantitative Analysis Using STEM-EELS/EDX Spectroscopy Techniques

2.1. Introduction

2.1.1. *Dopant analysis challenges in the silicon industry*

As mentioned in the general introduction of this book, the MOS transistor is the key component driving the electronic logic revolution for the last 50 years. However, strong problems can be predicted for MOS with a length less than 20 nm (i.e. four elemental silicon cells). Among the various phenomena, susceptible to limit the MOS, are the instabilities due to the statistical fluctuation of the number of dopant atoms inside small devices [BOI 88, PAC 00, HOE 72, THO 98]. This consideration is of primary importance because, for each new technology node, the definition of the doping architecture is one of the most challenging steps [DEN 74]. For the doped areas optimization, technologists have benefited from a previous knowledge in ions implantation, induced defects distribution, dopant diffusion, interaction with defects, furnace and rapid thermal annealing processes, recrystallization, dopant activation, simulation and electrical tests. Finally, complex doping architecture were always successfully developed pushing the limits of the MOS performances to their maximum. For each generation of technology, the CMOS transistors are smaller but give better switching performances in term of speed and I-on (I-off) currents. During the early developments, it was not possible to

---

Chapter written by Roland PANTEL and Germain SERVANTON.

detect at the nanometer-scale the dopant atoms and quantify their distribution. At that time this was not a problem because despite the lack of imaging techniques, doping architectures were successfully developed. However, for many years there has been a growing demand for 2D/3D dopant visualization and quantification by microcopy or imaging techniques. The reason is that all the physical phenomena involved during the present doping processes could not be predicted only by simulations and electrical measurements. In particular, this is the case for dopant segregation. Segregation can occur at the silicon/oxide interfaces of the gate, at the surfaces of silicon active areas and crystal defects, at the grain boundaries in silicon polycrystalline, and at silicide/silicon interfaces. Furthermore, during implantation, some areas are masked by material layers whose density and related stopping power can vary depending on others process steps. This is not easy to predict using simulations. Novel annealing processes were recently tested, but rapid lamp flash, laser annealing and the related physics mechanisms are not very well known. Also, the ion implantation and annealing processes, which are the easiest to simulate, are more and more combined or in competition with in situ doped epitaxy processes. This is the case for heteroepitaxial bipolar transistors (HBTs) and also for elevated source drain epitaxy of CMOS on bulk substrate or on silicon on oxide (SOI). Concerning the epitaxy processes that use low-pressure chemical vapor deposition (LPCVD) the dopant incorporation is extremely dependant on the gas mixture, the temperature, the crystal defects or the orientation of the growing facets. The resulting dopant distribution cannot be predicted by simulation and in this case physical characterization is mandatory. All the reasons mentioned above show that the simulation is blind to a lot of phenomena involved during the doping processes, and as a consequence the experimental detection and quantification of dopant distributions becomes more and more necessary. Furthermore, the transistors optimization of sub-30 nm MOS will be extremely sharp with very narrow process windows. In the following section, we evaluate briefly the different techniques presently available for dopant depth profiling (1D quantification) and for 2D or 3D quantitative mapping distribution.

2.1.2. *The different dopant quantification and imaging methods*

2.1.2.1. *Depth profiling methods over large areas*

As a very low concentration of doping atoms in silicon substrate can significantly change the electrical properties, in particular, the conductivity or p-n junction location, the ideal characterization techniques should be sensitive enough to detect concentration as low as 10^{15} cm^{-3}. The depth profiling techniques have this advantage: a high sensitivity and nanometer depth resolution but only inside large structures, that is with poor lateral spatial resolution. This is the case for the secondary ion mass spectrometry technique (SIMS) [WIL 89] and time-of-flight

SIMS (TOF-SIMS). In the microelectronic industry, these are the reference techniques being used from the beginning to study the physics of the doping process or control in line the implanted doses. Dedicated large boxes are included in the patterned silicon wafers to allow SIMS characterization. Inside nanometer devices, these techniques are not applicable and others solutions, less sensitive, but spatially resolved are presently actively tested.

2.1.2.2. Near-field microscopy techniques

Historically, the most used and sensitive technique to localize the exact electric p-n junction depth was the spreading resistance profiling (SRP) [HAR 98] method, which consists of measuring the resistance between two tips scanned over a low angle polished bevel of the junction. The modern version uses the tip of atomic force microscopes (AFMs) scanned on a cleaved cross-section of the device to obtain a 2D map. The tip measures either the resistance (scanning spreading resistance SSRM) [SUN 05, YAN 06, EYB 09, ZHA 10] or the capacitance (scanning capacitance SCM) [DUH 02, CIA 03, LIG 09]. The obtained map reflects the local carriers (electrons, holes) density, the p-n junction localization and active dopant concentration. These techniques present good sensitivity (10^{17} cm^{-3}) and high spatial resolution (few nanometers) but are not currently used because of the complexity of the overall experiment. At present, the near-field techniques are competitive for large- and low-doped areas in silicon substrate observation (isolation wells, pixels of CMOS imagers) but not really for MOS transistor analysis.

2.1.2.3. The atom probe tomography technique

The atom probe tomography (APT) technique was developed to study metal alloys 20 years ago [BLA 93a, BLA 93b]. A tip representing the sample to study is obtained by electro-erosion and placed in a high electrical field. Above a critical field, the atom of the tip surface are extracted, accelerated and detected on a 2D screen, which measures the *x,y* impact position and *t* the time of impact. If the field is pulsed, the atom mass can be determined by their time of flight. The observed object can be reconstructed in 3D assuming a knowledge of the trajectories of the detected atoms. At present, a pulsed laser is combined to a constant electric field and semiconductors or insulators materials can be studied [GAU 06a, GAU 06b, THO 06, VEL 06, GIL 07, KEL 07]. Using a focused ion beam (FIB) to etch tips exactly localized at a site-specific place, the APT technique is presently applied to study silicon devices. Three-dimensional distribution of all the principals atoms used in doping processes (P, As, B, C) can be obtained with atomic resolution [THO 07, DUG 09, DUG 10]. The sensitivity is in the range of 10^{18}–10^{19} cm^{-3}. Unequaled results have been obtained using APT, in particular concerning the segregation of boron [THO 05, CAD 09]. However, the technique has the following limitations: small analyzed volumes and difficult for certain materials, such as Si_3N_4 or the high K metal stack of gate of sub-30-nm technology.

2.1.2.4. *The transmission electron microscopy (TEM) techniques for dopant observation*

2.1.2.4.1. TEM classical bright field (BF) imaging

The TEM BF is one of the best imaging techniques when spatial resolution is concerned, 2A for classical TEM and sub-ångström in aberration corrected microscope. However, the physics of the TEM contrast, mainly governed by diffraction effects in crystal and mass thickness in amorphous material is by far not specific enough and sensitive to low-concentration impurities or dopant. In particular cases, the energy filtering (EFTEM) imaging technique can reveal chemical contrast but with a sensitivity of approximately 1%, that is not enough sensitive for dopant analysis.

2.1.2.4.2. TEM off axis electron holography

In TEM BF imaging, only the wave intensity, but not the phase, is measured. Therefore, interesting information is lost, in particular the phase change induced by the potential of p-n junctions. Rau proposed in 1999 [RAU 99] the TEM off-axis holography solution to visualize and quantify the internal potential of junctions. Using a dedicated TEM equipped with a post-sample mini-lens and a biprism, the electron wave is split into two parts, one traveling out of the sample and the other across it. Recombining these two wave parts gives an interference (holography) pattern from which the wave phase can be measured. This technique has been extensively applied for the 10 years [TWI 03, COO 09, COO 11, GRI 11]. It gives the internal potential map and coupled with simulation allow to estimate active dopant concentration distribution. The limitation is mainly due to the sample thickness (approximately 300 nm) and the necessity to tilt off axis (3°). Therefore, the spatial resolution is not competitive for 30 nm or sub-30-nm devices. On a large structure the sensitivity can be as good as 10^{17} cm^{-3}. This technique is discussed in detail in Chapter 1 of this book.

2.1.2.4.3. The analytical STEM techniques

2.1.2.4.3.1. STEM high angle annular dark field imaging (HAADF)

For heavy atoms incorporated in single silicon crystal, a Z contrast can be observed using large angle electron collection. The differential cross-section of elastic electron scattering by the atomic nucleus tends to follow the Rutherford law proportional to Z^2 at high angle (Z is the atomic number). This effect has been exploited to measure arsenic distribution in silicon using STEM HAADF imaging [MAC 93, WAN 01, FER 06, PEI 08, PAR 08]. However, the STEM Z contrast is sensitive to many parameters, crystal defects and sample thickness, and is therefore very hazardous or qualitative and limited to approximately 0.5% in the best case. This technique is presented and discussed in Chapter 7 of this book.

2.1.2.4.3.2. STEM analytical spectroscopy techniques (EELS and EDX)

The less ambiguous measure proving the presence of a particular atomic element inside a sample is the detection of a signal, which is a unique signature of this element. This is the aim of the spectroscopy techniques where the collected signal is due to electron interband or core-level transitions. In electron transmission microscopy, two physical phenomena can be exploited for spectroscopy applications: the electron energy loss (EELS) measured by electron spectrometers and the energy of X-rays emitted by the sample measured by dedicated detectors (energy dispersive X-ray spectroscopy EDS or EDX). In the STEM mode, scanned maps are collected pixel-by-pixel to construct "spectrum images" [JEA 89, COL 94, MUL 08]. The data cube has three dimensions: two spatial and one energy (x, y, E). Only recently these spectroscopy techniques have been applied for the quantification of dopant distribution in silicon devices. EELS was used for arsenic dopant analysis [SER 10a, SER 10b, SER 12, COO 11, TOP 03]. A previous STEM-EDX study [TOP 03] of MOS transistors using Si(Li) EDX detectors were significantly improved recently with the use of powerful silicon drift detectors (SDDs) in TEM [PAN 11, PAN 12, CLE 11]. In particular, arsenic and phosphorus can now be analyzed with a detection limit in the range of 10^{19} cm^{-3} and with approximately 1 nm spatial resolution. A complete presentation of the issues concerning instrumentation, methods, application example and performance evaluation of STEM-EELS-EDX quantitative dopant mapping is described in the following sections.

2.2. STEM-EELS-EDX experimental challenges for quantitative dopant distribution analysis

2.2.1. *Instrumentation present state-of-the-art and future challenges*

2.2.1.1. *Electron sources and probe aberration correctors*

For the optimization of the STEM-EDX-EELS performances, mainly spatial resolution and detection limit, the electron source should provide a maximum current in a smallest probe. The electron source should have preferably a high brightness, a small real or virtual size, a high maximal current above few nanoamperes, high stability and low noise. The cold field emission gun (FEG) respects the first two criteria but less the other ones (stability, noise and max current). The Schottky FEG are less powerful in brightness and source size but are stable and give a high current. A recent improvement is the high brightness source Schottky (X-FEG) [FRE 08], which approaches all the criteria. The probe size is in fact determined by the spherical aberration of the lens forming it. A more costly and complex solution is therefore a spherical aberration corrector probe, which allows us to focus a larger convergent angle beam increasing the current while keeping the resolution near 1 Å. For the quality of STEM analytical experiments, electronic noise of the scanning unit and acoustic vibration of column and sample holder, should also be minimized.

2.2.1.2. Electrons and X-rays energy dispersive spectrometer and detectors

In analytical STEM, the spectrometers and detectors performances determine the signal quality and detection sensitivity limits. The most important characteristic is the percent of the collected signal compared to the total available one which should be maximum. The latest Gatan EELS spectrometer models have improved the collection angle by using s large entrance aperture (5 mm) and sophisticated spectrum focusing with added lenses. In the corrected NION STEM probe a correction system placed at the spectrometer entrance improves the EELS collection up to 100 mrad [KRI 08]. The X-ray detectors collection angles have also been improved by decreasing the sample detector distance in the TEM and also placing many detectors (four in Osiris). The detection solid angle has increased from 0.15 to 0.9 sr. The other important characteristics of the spectrometer are the number of channels, the energy resolution, the reading speed and the maximal number of counts per second. For a long time, EELS spectrometers, despite their progressive improvement, have suffered energy range limitation (1,000 eV), too slow CCD detector reading speed (approximately 10–100 ms per spectrum) and insufficient signal dynamic collection. This has limited the acceptable number of pixels in spectrum images. The recently developed new spectrometer Quantum Dual EELS from Gatan [THO 10] seems to have solved these problems, giving access to the total spectrum energy range (0.0–3,000 eV) with dynamic allowing simultaneous low loss and core loss acquisition and with improved reading speed (2 ms per spectrum). In EDX analysis, a revolution was introduced by the SDDs, developed first for spatial applications. For nearly 10 years the SDDs have been preferred for installment in SEM replacing the Si(Li) diode X-ray detectors, and they are also now used more in TEM/STEM equipment [FAL 10, VON 09, SCH 10]. These devices show very high counting rates up to one million photons per second, improved energy resolution and can be run at room temperature. With the electron source and detectors improvements, STEM spectroscopy dopant analyses carried out on 10-year-old equipment [SER 10a] will probably be improved and will stay competitive compared to other techniques for application on devices of future silicon nodes (20 nm).

2.3. Experimental conditions for STEM spectroscopy impurity detection

2.3.1. Radiation damages

The collected signal resulting from the interaction of the electron beam on a particular area of a sample (say a pixel in a spectrum image) will be proportional to the electron dose; that is the product of beam current by the exposure time (pixel dwell time). Therefore, the signal can always be increased by the current and the exposure time. The only limitation remains that the sample should support the local radiation without being modified (or more realistically with acceptable limits). Studies on silicon crystal [EGE 04] have shown that the main radiation damage process is the knock-on effect where the electron momentum is transmitted to a silicon atom. The maximum energy displacement, E_{tmax}, of a species with atomic mass m_{Si}, hit by incident electron with energy E_i, is given by the following formula [BIL 61]:

$$E_{t_{max}} = \frac{2(E_i + 2m_e c^2)}{m_{Si} c^2} E_i,$$

with m_e the electron mass and c the speed of light.

An electron accelerated at 200 keV primary energy can transfer 20 eV to Si atoms, which are then displaced because the estimated displacement energy has experimentally been evaluated to be 13 eV [RAP 55]. Radiation damages are then decreased by lowering the primary energy because the maximum momentum transfer decreases. Figure 2.1 shows the evolution versus time of the Si-Ka EDX signal emitted by a 80-nm-thick silicon sample impacted by focused electron probe (2.5 nA) at different primary beam energies (200, 160, 120 and 80 keV). The EDX signals decrease because the Si atoms are locally sputtered, which leads to a local hole in the lamella. At 200 and 160 keV, the radiation damages are important. For lower energies, 120 keV and less, the radiation effects become negligible (or acceptable) and we can expect even better sample hardness at 80 keV.

Lowering radiation damages when performing spectroscopy is crucial because structural defects such as dislocation, grain boundaries or stacking faults are susceptible to enhance dopant diffusion when radiation damages increase [HOB 79]. Furthermore, the ionizations cross-sections at origin of the spectroscopic signals increase when the primary energy is lowered. Finally, for silicon dopant analysis, 80 keV could be a good compromise if a X-FEG source is used, the current can be increased above 4 nA, and spatial resolution as good as 2 nanometer.

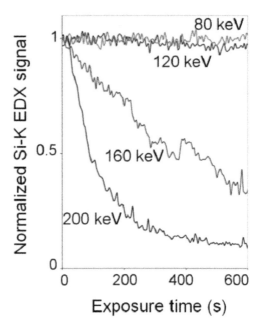

Figure 2.1. *Evolution of the Si radiation damages kinetics as a function of the exposure time and for several primary beam energies*

2.3.2. *Particularities of EELS and EDX spectroscopy techniques*

2.3.2.1. *Electron Energy Loss Spectroscopy (EELS)*

The mechanisms at the origin of the primary electrons energy losses are the inelastic interaction with electron populations in the sample (collective plasmon interaction or direct ionization of core levels). The plasmon interaction does not give specific element identification but gives other information such as the sample thickness. The core loss signatures give access to the detection and quantification of all atomic elements in the sample. The energy loss is measured by an electron spectrometer. The experimental energy resolution is defined by the source chromaticity (approximately 0.7 eV for a Schottky FEG). The electron energy loss spectra are collected by a CCD camera in which vertical pixel rows are summed to construct the energy channels. In a typical EELS spectrum the background is high compared to the ionization edges core loss structures and generally a power law fits well the background that can be extrapolated below the core loss edges. The spectrum can be quantified after background extrapolation using a fit of the core loss edges by theoretical cross-sections. The EELS technique coupled with STEM imaging spectrum mode, was recently applied to the analysis of arsenic dopant distribution in silicon [SER 09, SER 10a, SER 10b, SER 12]. The EELS As-$L_{2,3}$ ionization threshold energy appears at 1,323 eV loss. This edge is superposed to a high background, which is subtracted by fitting a power law (E^{-r}). Figure 2.2 shows a typical EEL spectrum from a highly doped As region (3.7%) presenting the As-$L_{2,3}$, As-L_1 and Si-K edges before and after background extrapolation. The electron dose used for the acquisition of such a spectra is (0.8 nA × 10 s).

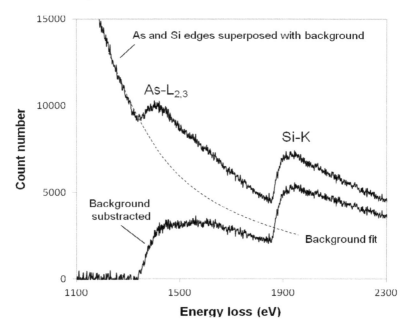

Figure 2.2. *120 keV As-$L_{2,3}$ EELS spectrum and background removed from a 3.7% (18.5 × 10^{20} cm^{-3}) As-doped Si layer*

The background extraction window is fixed to 150 eV width [1,173 eV; 1,323 eV] and the As edge integration window is fixed to 260 eV width [1,323 eV; 1,583 eV]. Concerning Si-K edge, which appears at 1,838 eV, the background extraction and edge integration window were both fixed to 150 eV widths. Furthermore, in order to reduce the beam radiation damages and enhance the ionization cross-section, the acceleration voltage was lowered from 200 keV to 120 keV [SER 10a, SER 12].

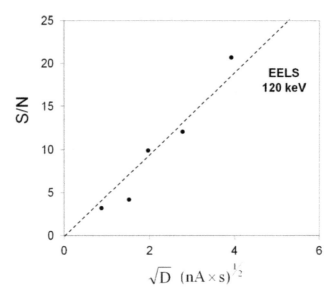

Figure 2.3. *Evolution of the EELS As-L signal-to-noise ratio measured at 120 keV on a 0.7% arsenic-doped silicon and plotted versus $\sqrt{D}$*

The signal-to-noise ratio (S/N) versus the electron dose has been tested on STEM EELS at 120 keV and is shown in Figure 2.3. This shows that the STEM EELS S/N evolution versus $\sqrt{D}$ on a 0.7% As-doped Si (As:Si). The As signal (S) has been calculated with a background extraction and integration windows of 150 eV and 260 eV, respectively. The total noise N is the addition of As signal noise plus Si background noise. Figure 2.3 shows that the STEM EELS signal-to-noise ratio S/N is linear with $\sqrt{D}$, which is expected from statistics. STEM EELS S/N value reaches 20 for an electron dose $\sqrt{D} = 4$ $(nA.s)^{1/2}$, which means for a 0.7% As:Si that the arsenic detection limit using this technique for such electron dose is close to 0.04% (2×10^{19} cm^{-3}).

2.3.2.2. Energy dispersive X-ray spectroscopy (EDS-EDX)

The atoms of a sample irradiated by energetic electrons can be excited or ionized with a certain probability. This probability is governed by differential cross-sections depending on the electron primary energy, the energy transferred and the scattering angle. The excited or ionized atoms recover their fundamental-level state by emitting radiations, mainly Auger electrons or X-rays. The X-rays are due to direct interband electrons recombination and they transport the energy difference between the initial and final states. The resulting series of narrow X-ray peaks at discrete energies give a clear signature of the atomic elements in the sample. The energy dispersive X-ray spectroscopy (EDS-EDX) technique is the simultaneous collection and energy measurement of X-ray photons. It uses dedicated detectors: Si(Li) diodes and more frequently today SDDs. The EDX spectra contain narrow Gaussian peaks (100–200 eV width) superposed to a nearly horizontal background due to the bremsstrahlung effect. Figure 2.4 shows experimental examples of EDX spectra obtained on a silicon sample phosphorus and arsenic doped. A basic signal extraction consists in the collection of photons inside a fixed energy window centered on the chosen peak. However, this method mixes the signal and the background. A better method is the spectrum quantification, which combines EDX peak fitting using tabulated Gaussians functions and background removal. Theoretical sensitivity factors are used for the elemental quantification, and minor corrections can be added using K factors. For low-energy X-ray peaks and a heavy element sample matrix, absorption correction must be included.

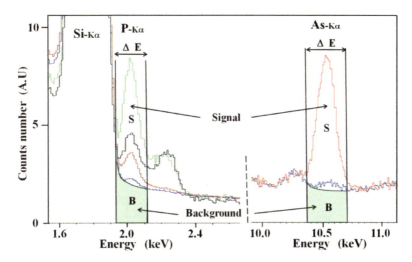

Figure 2.4. *EDX signal extraction on phosphorus-doped (0.08–1.3%) or arsenic-doped (0.25%) silicon TEM sample. Only spectrum areas around P-Kα and As-Kα are shown. The extraction window ΔE, signal S and background B are shown*

A spectrum map extracted from the 3D data cube (x, y, E) contains the signal (S) and the background (B) as shown in Figure 2.4. Therefore, the fluctuation of counts number in each pixel of the elemental map is the fluctuation of S plus the fluctuation of B. For a simple model to estimate the signal-to-noise ratio over noise ratio we will assume a high number of counts for both S and B and, therefore, a Gaussian statistic behavior. We will also assume that the fluctuations stay inside a standard deviation (68% probability) and therefore we can calculate the signal over noise S/N, using noise $N = S^{1/2} + B^{1/2}$. The signal and background are proportional to the electron dose impinging on a defined area of the sample (pixel or local area). Concerning the EDX signal S of diluted impurities in silicon we can write:

$$S = a_s \, x \, t \, D \tag{2.1}$$

S is the signal of the impurity (number of counts), a_s is a constant or cross-section, x is the impurity concentration (%), t is the sample thickness (nm) and D is the electron dose (nA.s). For the background B under the signal S and due to the silicon matrix (bremsstrahlung) we have also the t and D proportionality:

$$B = a_b \, t \, D \tag{2.2}$$

So we can calculate S/N

$$S/N = \{(a_s \, x - a_b)/[(a_s \, x)^{1/2} + a_b^{1/2}]\} \, (t \, D)^{1/2} \tag{2.3}$$

The expressions above show that the signal and background follow (tD), whereas the signal over noise follow $(tD)^{1/2}$.

Figure 2.5 shows an experimentally typical example, showing the EDX signal over noise ratio versus the square root of the electron dose obtained using a Tecnai Osiris at 80 keV energy. The samples are 100-nm-thick silicon lamella arsenic or phosphorus doped (0.2%). If we consider that the dopant is detectable for a signal over noise ratio superior to 1, for arsenic, 0.02% (1×10^{19} cm^{-3}) is detectable using a 4 nA × 2.5 s dose. In the case of phosphorus, the same level detection is also possible using 4 nA × 6 s dose.

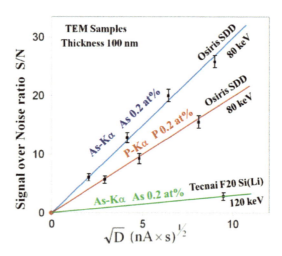

Figure 2.5. *Experimental EDX signal over noise ratio versus the square root of the electron dose obtained using a Tecnai Osiris at 80 keV energy. The samples are 100-nm-thick silicon lamella arsenic or phosphorus doped (0.2%)*

These doses are acceptable at 80 keV and 120 keV because, as explained in section 2.3.1, there is no noticeable radiation damages at such energies. For comparison, in Figure 2.5 we have also plotted the arsenic signal over noise ratio that would be obtained on a similar sample (0.2% As) at 120 keV on a Tecnai F20 equipped with a Li(Si) diode [SER 09]. It is shown that 0.02% would demand a 1 nA × 100 s dose knowing that for saturation reasons the current is limited to 1 nA.

2.3.3. *Equipments used for the STEM-EELS-EDX analyses presented in this chapter*

The first EELS quantitative arsenic distribution analysis in silicon [SER 10a, SER 12] was carried out using a Tecnai F20 equipped with a Gatan GIF 200. The microscope and electron spectrometer were aligned at 120 keV energy used for this study. The beam current was approximately 5 nA. STEM EELS spectrum imaging acquisitions were carried out during sequences of approximately 2 h, using active drift correction and collecting approximately 8,000 pixels. The signal extraction and quantification of As:Si is presented in [SER 09, SER 10a]. The results and discussion are detailed in section 2.4.1.

Preliminary STEM EDX experiments using a Tecnai F20 equipped with Si(Li) diodes have shown poor sensitivity compared with EELS [SER 09]. The installation of a Tecnai Osiris in our laboratory has improved the EDX signal (for more than two decades) such that only dopant analysis using Osiris will be presented. The Osiris TEM is equipped with a X-FEG electron source providing, compared to a Schottky FEG, a higher current (factors of three to five) for a same probe size. The EDX detection system included four SDDs placed in the pole piece totalizing 0.9 sr solid angle collection. Fast reading and high detection rate allow us to collect few 10^5 X-rays per second with a low dead-time (5%). Note that the Osiris also included a Gatan Enfina EELS spectrometer and that EELS-EDX signal can be acquired at the same time. For all the applications presented in this chapter, the samples were prepared and thinned using FIB at 30 keV with a final cleaning at 5 keV. The samples thickness is approximately 80–100 nm.

2.4. STEM EELS-EDX quantification of dopant distribution application examples

2.4.1. *EELS application analysis examples*

2.4.1.1. *Heterojunction bipolar transistors*

The first application on HBT was carried out using a Tecnai FEG F20 equipped with GIF 200. Bipolar transistors are high-frequency analogical components used in today's telecommunications (wireless, LAN) and automotive (radar) applications.

When coupled with CMOS transistors, Bipolar are named BiCMOS transistor and can treat both analogical and logical signals. A TEM BF image shows a similar BiCMOS transistor on Figure 2.6. The device is made of three main region: an As:Si emitter part, a Ge:Si Base and a collector region. The As:Si emitter doping is performed using LPCVD with SiH_4 and AsH_3 flux gases. The homogeneous dopant distribution inside such devices is crucial for an optimized electrical behavior. As explained in section 2.1.2.4.1 and shown in Figure 2.6, the TEM BF contrast is too weak and mixed with diffraction contrast. Therefore, it is not possible to evaluate the presence of arsenic dopant atoms inside the emitter region. Using the STEM EELS spectrum-images technique, we demonstrate the direct detection of arsenic dopant inside a semiconductor device, as shown in Figure 2.7. The STEM EELS map acquired at 120 keV acceleration voltage on a FEI TECNAI microscope, during 2 h for 150 × 50 pixels, reveals strong As distribution heterogeneities. High As concentrations are detected at the BiCMOS emitter edge (2% As:Si detected at Si_3N_4 edges) whereas the emitter center is partially depleted (~0.5% As:Si). Dopant segregation with As concentrations up to 2% at poly-silicon grain boundaries are also evident. Dopant incorporation issues, which are key studies for process engineers, can now be controlled using such spectroscopic technique in order to monitor and optimize device performances.

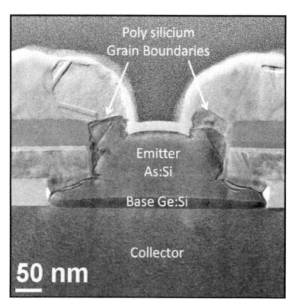

Figure 2.6. *TEM bright field image of n–p–n 90 nm BiCMOS transistor*

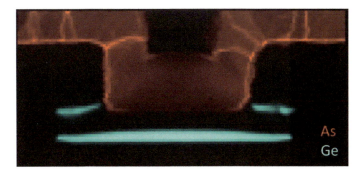

Figure 2.7. *Quantified STEM EELS As (red) and Ge (blue) map obtained on the same BiCMOS than Figure 2.6. The As concentration in the part of emitter is inferior to 1%, the Ge concentration is near 20%*

Figure 2.8 shows a TEM BF of another BiCMOS transistor with a grain boundary revealed by BF contrast between the center emitter part (monocrystalline) and the emitter right edge (polycrystalline). The STEM EELS map from Figure 2.9 shows the As dopant distribution using a "temperature" false color, which is proportional to the As percentage. The atomic concentration ratio between two elements As and Si is linked to the measured signal intensities ratio (I_{As}, I_{Si}) by the Cliff–Lorimer equation [CLI 75]:

$$C_{As}/C_{Si} = k_{As}\, I_{As}/I_{Si}$$

with k_{As} a sensitivity factor, which depends on various experimental factors (acceleration voltage, lamella thickness). Details concerning As dopant quantification method using STEM EELS can be found in [SER 09]. The STEM EELS map presented in Figure 2.9 has been acquired at 120 keV during 2.2 h for 150 × 60 pixels. Dopant concentration variations are detected inside the emitter. Furthermore, the grain boundary seems to act as a dopant trap. It is also likely that arsenic atoms incorporate poorly in the silicon emitter part during the epitaxy process and diffuse to the grain boundaries and Si_3N_4 spacers edges after thermal annealing.

In conclusion, the quantitative arsenic dopant maps using STEM EELS at 120 keV enable to analyze dopant distribution in state-of-the-art semiconductor devices compatible with standard FIB lamella preparation. The spatial resolution is 2 nm and sensitivity in the low 10^{19} cm^{-3}. The TEM lamellas were prepared using FIB milling with a final cleaning stage at 5 keV to lower Si amorphization. Details concerning STEM EELS technique optimization for As distribution maps and corresponding applications in semiconductor nanodevices can be found in [SER 09, SER 10a].

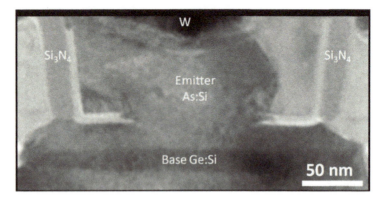

Figure 2.8. *TEM bright field image of n–p–n 90 nm BiCMOS transistor*

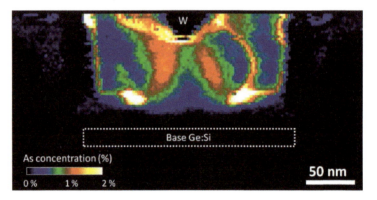

Figure 2.9. *Quantified STEM EELS As map obtained on the same BiCMOS as Figure 2.8. The display is in color temperature and the green-red limit is approximately 1%*

2.4.1.2. *NMOS high-K metal gate 28 nm technology analysis*

The second application on a high-K metal gate *n*MOS transistor of advanced technology (28 nm) was carried out on the Tecnai Osiris combining X-FEG and Gatan Enfina EELS spectrometer. The raw extraction of arsenic quantitative map as explained in section 2.3.2.1, is shown in Figure 2.10. The map shows arsenic distribution in source, drain and gate. These EELS data series were sampled over 180 × 65 pixels on FEI TEM Osiris aligned at 120 keV acceleration voltage. The 12,000 spectra are acquired with a detector reading speed of 0.3 s per spectrum leading to an overall mapping time of approximately 1 h. Segregation is observed all along the external surfaces of the gate and also near the top surface of source (drain). The interesting area is the frontier of doped source (drain) defining the p-n junction limit and this frontier is a little bit noisy.

Dopant Distribution Quantitative Analysis 53

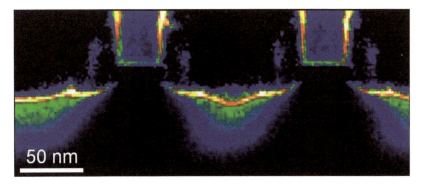

Figure 2.10. *Quantitative EELS map of arsenic distribution in a 28 nm technology NMOS transistor. The display is in color temperature scale and the green/red limit is approximately 1%*

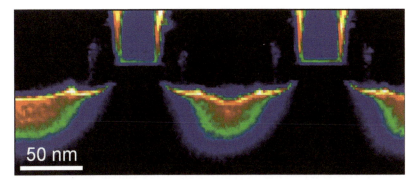

Figure 2.11. *Reconstructed quantitative arsenic distribution map using multivariate statistical analysis of the STEM-EELS spectrum image for same original data shown in Figure 2.10. The displayed principal components analysis corresponding to arsenic is used to fit the data*

Spectrum images containing a lot of data can be treated using more advanced methods, especially using multivariate statistical analysis (MSA), developed 10 years ago. In particular, the principal component analysis (PCA) [TIT 96, TIT 99, KOT 06] or independent component analysis (ICA) [BON 05, DEL 10] are efficient methods able to extract the significant signal from background and noise. Figure 2.11 shows a similar quantitative arsenic map as in Figure 2.10 but treated using a PCA method and fitting with the arsenic principal component. Figure 2.11

shows an improvement of signal over noise ratio, especially at the frontier of source/substrate or drain/substrate p-n junctions.

2.4.2. *EDX application analysis examples*

2.4.2.1. *Dopant/impurity segregation analysis*

The segregation of impurities, and in particular of atoms introduced in silicon to change the electric properties, is an important effect that should not be neglected or ignored because it could be the origin of statistical fluctuation or device anomalies. Segregation can occur at silicon/silicon oxide interfaces [TOP 03], at grain boundaries and at defects in the silicon crystal. Generally the segregated dopant atoms are not activated because they are not in normal substitutional sites or are bounded to non-silicon atoms. Therefore, the segregation can change significantly the concentration of active dopant. Unfortunately, for a long time no techniques spatially resolved and sensitive enough were available to directly measure the dopant segregation. At present, the APT technique has proven to be the ultimate solution for segregation analysis of B, P and As [THO 05] but this method is complex and is restricted to small volumes (50–100 nm diameter). The application shown below evaluates the possibility given by the STEM-EDX technique on a modern TEM microscope (Osiris) using combining X-FEG and multi SDDs. Figures 2.12(a)–(c) show STEM-EDX analysis of silicided ($CoSi_2$) silicon polycrystalline gate doped with arsenic and phosphorous. On the STEM image (Figure 2.12(a)) the crystal contrast allows us to estimate grain size and grain boundaries. The Figures 2.12(b) and 2.12(c) show the distribution of Co, As, P and O elements. This analysis was obtained using 4 nA current at 120 keV energy, 1,000 × 300 pixels for a total exposure time of 1 h in Osiris TEM. Clearly, segregation of arsenic and phosphorus is observed at grain boundaries. Also, non-homogeneous distribution is observed in the CoSi2 inside or at external surfaces and at the ONO top surface. The STEM-EDX maps are a 2D projection of the 3D structure and the grain boundaries nearly parallel to the electron beam are the most visible ones. A global observation shows that the phosphorus concentration is higher in the low part of poly-silicon and, on the contrary, the arsenic is more concentrated in the top part. This is due to the different implantation energy and the higher phosphorus diffusion coefficient.

Dopant Distribution Quantitative Analysis 55

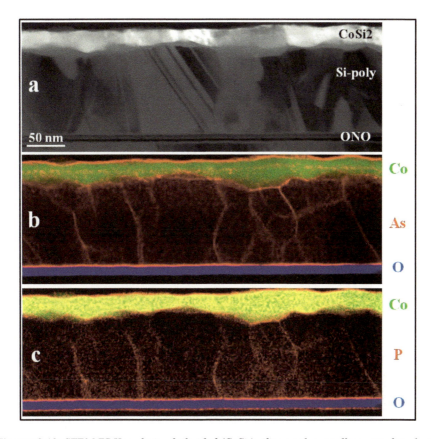

Figures 2.12. *STEM-EDX analysis of silicided (CoSi$_2$) silicon polycrystalline gate doped with arsenic and phosphorous. a) STEM HAADF Z contrast image showing crystal contrast in individual grains. b) STEM-EDX elemental distribution of Co, As, O.*
c) STEM EDX elemental distribution of Co, P, O

Figure 2.13 shows a zoomed view of a grain boundary and a line profile at a place where the segregation of arsenic is maximum (grain boundary aligned to the electron beam). The As concentrations in the grains are approximately 0.1–0.2% and on the grain boundary it rises to 0.8%. This is convoluted with the experimental spatial resolution, which is less than 2 nm because the width of As profile is 2 nm. A 3D analysis should be ideally necessary to quantify segregation at grain boundaries. The atom probe technique is unique for that capability; however, STEM EDX is a more easy experimental method, a large field of view can be obtained and complete or partial tomography is possible by a series of using different angle analysis.

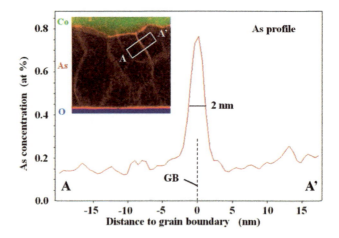

Figure 2.13. *Line profile of the arsenic concentration at a grain boundary. Upper left: zoomed view of this grain boundary and indication of the place of the line profile*

2.4.2.2. *Dopant distribution in a 40-nm NMOS transistor*

As mentioned in the introduction of this chapter, direct observation and quantification of the dopant atoms distribution inside devices such as a MOS transistor is demanded more and more in order to optimize the fabrication process. Figures 2.14(a) and (b) show STEM-EDX elemental maps distribution of arsenic and oxygen (Figure 2.14(a)) and phosphorus and oxygen (Figure 2.14(b)) obtained on a cross-section of a 40-nm NMOS transistor at the fabrication step before silicide formation. The two deep areas doped in silicon substrates are the source/drains. Low doped extensions (LDD) near the gate and near the substrate surface define the MOS transistor channel length. At the center, the gate is lowly doped but parts of the dopant atoms have segregated on all external surfaces. At each side of the gate, the spacers have protected the LDD areas during the high energy source/drain implantation. Therefore, arsenic and phosphorus are present in theses spacers. All these information are valuable to control the quality process, in particular the verification of the stopping power of spacers is not easily predicted by simulations. The measure of the real concentration in the gate is also valuable because part of the active dopant atoms is lost by segregation at external surfaces. The exact measure of the dopant extension and channel length is also interesting because it can be used to refine electrical measurements and simulations. Figures 2.15(a) and 2.15(b) show zoomed STEM-EDX maps of the channel area: arsenic (Figure 2.15(a)) and phosphorus (Figure 2.15(b)) distributions. The display is a quantitative temperature scale and the green/red limit is equal to 1%. The pixel size of the STEM-EDX maps is equal to 2 Å. From Figures 2.15(a) and (b), it is evident that the arsenic atoms in the LDD areas extend nearer the gate than the phosphorus atoms.

Dopant Distribution Quantitative Analysis 57

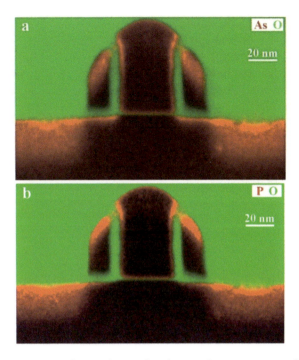

Figures 2.14. *STEM-EDX elemental maps distribution of a cross-section of 40 nm NMOS transistor: a) arsenic and oxygen, b) phosphorus and oxygen*

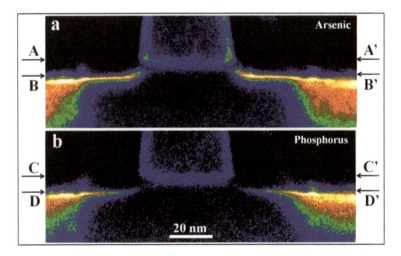

Figure 2.15. *Zoomed STEM-EDX maps of the NMOS channel area: a) arsenic and b) phosphorus quantitative distributions. The display is a quantitative temperature scale and the green/red limit is equal to 1%*

Quantitative line profiles along lines defined in Figures 2.15(a) and 2.15(b) are shown in Figures 1.16(a) and 1.16(b). The gate lengths measured in arsenic (AA′) and phosphorus (CC′) profiles are found equal to 43 nm. The LDD extensions are different and the channel length defined by arsenic (BB′ profile) is 32 nm whereas it is equal to 40 nm for phosphorus (DD′ profile). The electrical channel length is therefore governed by the limit of arsenic extension and is equal to 32 nm. These STEM-EDX experiments were carried out on Osiris at 120 keV, 4 nA with approximately 1,000 × 500 pixels and approximately 1 h exposure time.

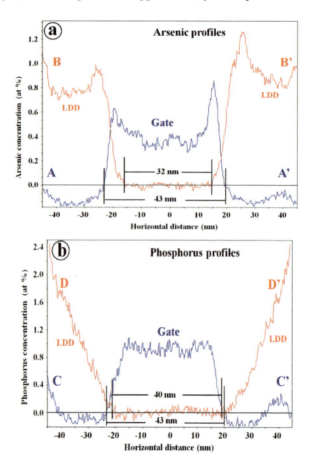

Figure 2.16. *Quantitative doping concentration profiles extracted along lines indicated in Figure 2.15. The gate and channel lengths defined by the As a) or P b) profiles are measured and indicated in the figures*

2.5. Discussion on the characteristics of STEM-EELS/EDX and data processing

The first EELS experiments was carried out a few years ago using a 10-year-old TEM (Tecnai F20) which gave good results for arsenic quantitative distribution mapping [SER 09, SER 10a, SER 10b, SER 12]. The sensitivity limit was estimated to be approximately 0.02% and the resolution approximately 2 nm. The quality of these results can be improved by modern microscopes using X-FEG, the last generation Enfina and better electronic and acoustic stability (see some new results in section 2.4.1). However, background extraction, easy in silicon using a power law, is more complex when dopants are distributed at proximity of silicon alloys interfaces (silicides or SiGe). Moreover, only arsenic analysis was proven using EELS. In the case of phosphorus and boron ionization cross-section and complex background are not favorable. Using a new powerful microscope equipped with EDX SDDs, STEM-EDX, analysis of dopant seems to be practicable with enough sensitivity. However, the case of boron is not solved. EDX data processing and quantification can be easier than EELS treatment but some time peaks interference or spectrum background could be difficult. Advanced data processing solutions such as MSA and PCA [TIT 99, KOT 06, TIT 96] are known to be powerful methods to extract information from image spectrum. These should be more systematically tested because they probably solve interference and background problems. However, the principal components should be always interpreted by the microscopist, who must verify the physical or chemical significance, which must verify the physical or chemical significance of such signatures. Also, more *ab initio* techniques, called independent components analysis (ICA), are developed and are suspected to be more powerful for EELS spectrum data extraction [BON 05, DEL 10].

2.6. Bibliography

[BIL 61] BILLINGTON D.S., CRAWFORD J.H., *Radiation Damage in Solids*, Princeton Press, Seattle, WA, 1961.

[BLA 93a] BLAVETTE D., DECONIHOUT B., BOSTEL A., SARRAU J.M, BOUET M., MENAND A., "The tomographic atom probe: a quantitative three-dimensional nanoanalytical instrument on an atomic scale", *Review of Scientific Instruments*, vol. 64, pp. 2911–2919, 1993.

[BLA 93b] BLAVETTE D., BOSTEL A., SARRAU J.M., DECONIHOUT B., MENAND A., "An atom probe for three-dimensional tomography", *Nature*, vol. 363, pp. 432–435, 1993.

[BOI 88] BOIS D., ROSENCHER E., "Les frontières physiques de la microélectronique", *La Recherche*, no. 203, pp. 1176–1186, October 1988.

[BON 05] BONNET N., NUZILLARD D., "Independent component analysis: a new possibility for analysing series of electron energy loss spectra", *Ultramicroscopy*, vol. 102, pp. 327–337, 2005.

[CAD 09] CADEL E., VURPILLOT F., LARDE R., DUGUAY S., DECONIHOUT B., "Depth resolution function of the laser assisted tomographic atom probe in the investigation of semiconductors", *Journal of Applied Physics*, vol. 106, pp. 044908–044908-6, 2009.

[CIA 03] CIAMPOLINI L., BERTIN F., HARTMANN J.M., ROCHAT N., HOLLIGER P., LAUGIER F., CHABLI A., *Materials Science and Engineering B*, vol. 102, pp.113–118, 2003.

[CLE 11] CLEMENT L., BOROWIAK C., GALAND R., LEPINAY K., LORUT F., PANTEL R., SERVANTON G., THOMAS R., VANNIER P., BICAIS N., "Microscopy needs for next generation devices characterization in the semiconductor industry", *Proceedings of the 17th International Conference on Microscopy of Semiconducting Materials, J. Phys. Conf. Ser.*, p. 326, 2011.

[CLI 75] CLIFF G., LORIMER G.W., "The quantitative analysis of thin specimens", *Journal of Microscopy*, vol. 103, pp. 203–207, 1975.

[COL 94] COLLIEX C., TENCÉ M., LEFÈVRE E., MORY C., GU H., BOUCHET D., JEANGUILLAUME C., "Electron energy loss mapping", *Mikrochimica Acta*, vol. 114/115, pp. 71–75, 1994.

[COO 09] COOPER D., RIVALLIN P., HARTMANN J.M., CHABLI A., DUNIN-BORKOWSKI R.E., "Extending the detection limit of dopants for focused ion beam prepared semiconductor specimens examined by off-axis electron holography", *Journal of Applied Physics*, vol. 106, 064506, 2009.

[COO 11] COOPER D., DE LA PENA F., BECHE A., ROUVIERE J.L., SERVANTON G., PANTEL R., "Field mapping with nanometer-scale resolution for the next generation of electronic devices", *Nano Letters*, vol. 11, no. 11, pp. 4585–4590, 2011.

[DE 10] DE LA PENA F., BERGER M.H., HOCHEPIED J.F., DYNYS F., STEFAN O., WALLS M., "An application of independent component analysis in EELS: mapping spinodally decomposed titanium and tin oxide phases", *17th International Microscopy Congress*, Rio, pp. I5–18, September 2010.

[DEN 74] DENNARD R.H., GAENSSLEN F.H., YU H-N., RIDEOUT V.L., BASSOUS E., LEBLANC A.R., "Design of Ion-Implanted MOSFET'S with very small physical dimensions," *IEE Journal of Solid State circuit*, vol. SC-9, no. 5, pp. 256–268, 1974.

[DUG 09] DUGUAY S., VURPILLOT F., PHILIPPE T., LARDE R., DECONIHOUT B., SERVENTON G., PANTEL R., "Evidence of atomic-scale arsenic clustering in highly doped silicon," *Journal of Applied Physics*, vol. 106, pp. 106102–106102-3, 2009.

[DUG 10] DUGUAY S., CADEL E., SERVANTON G., PANTEL R., Conférence EMRS Fall meeting, September 2010.

[DUH 02] DUHAYON N., CLARISSE T., EYBEN P., VANDERVORST W., HELLEMANS L., *Journal of Vacuum Science and Technology B*, vol. 20, pp. 741–746, 2002.

[EGE 04] EGERTON R.F., LI P., MALAC M., *Micron*, vol. 35, p. 399, 2004.

[EYB 09] EYBEN P., MODY J., NASIR A., SCHULZE A., HANTSCHEL T., VANDERVORST W., "Sub-nanometer two-dimensional carrier profiling in silicon MOS technologies using high vacuum scanning spreading resistance microscopy", *ECS Meeting Abstracts*, vol. 902, 2009.

[FAL 10] FALKE M., KROEMER R., FISSLER D., ROHDE M., "News on silicon drift detectors for X-ray nanoanalysis in S/TEM", *Microscopy and Microanalysis*, vol. 16, no. S2, pp. 2–3, 2010.

[FER 06] FERRI M., SOLMI S., PARISINI A., BERSANI M., GIUBERTONI D., BAROZZI M., "Arsenic uphill diffusion during shallow junction formation", *Journal of Applied Physics*, vol. 99, p. 113508, 2006.

[FRE 08] FREITAG B., KNIPPELS G., KUJAWA S., VAN DER STAM M., HUBERT D., TIEMEIJER P.C., KISIELOWSKI C., DENES P., MINOR A., DAHMEN U., "First performance measurements and application results of a new high brightness Schottky field emitter for HR-S/TEM at 80–300 kV acceleration voltage", *Microscopy and Microanalysis*, vol. 16, no. S2, p. 1370, 2008.

[GAU 06a] GAULT B., MENAND A., DE GEUSER F., DECONIHOUT B., DANOIX R., "Investigation of an oxide layer by femtosecond-laser-assisted atom probe tomography", *Applied Physics Letters*, vol. 88, 2006.

[GAU 06b] GAULT B., VURPILLOT F., GILBERT M., VELLA A., MENAND A., BLAVETTE D., DECONIHOUT B., "Design of a femtosecond laser assisted tomographic atom probe", *Review of Scientifics Instruments*, vol. 77, p. 043705, 2006.

[GIL 07] GILBERT M., VURPILLOT F., VELLA A., BERNAS H., DECONIHOUT B., "Some aspects of the silicon behaviour under femtosecond pulsed laser field evaporation," *Ultramicroscopy*, vol. 107, 2007.

[GRI 11] GRIBELYUK M.A., OLDIGES P., RONSHEIM P.A., YUAN J., KIMBALL L., "Suppression of boron diffusion in deep submicron devices," *Journal of Vacuum Science Technology*, vol. 29, no. 6, pp. 062201–062204, 2011.

[HAR 98] HARRINGTON W.L., MAGEE C.W., PAWLIK M., DOWNEY D.F., OSBURN C.M., FELCH S.B., "Techniques and applications of secondary ion mass spectrometry and spreading resistance profiling to measure ultrashallow junction implants down to 0.5 keV B and BF_2", *J. Vac. Sci. Technol. B*, vol. 16, no. 286, pp 286–291, 1998.

[HOB 79] HOBBS L.W., *Radiation Effects in Analysis of Inorganic Specimens by TEM, Introduction to Analytical Electron Microscopy*, Plenum Press, New York, pp. 437–480, 1979.

[HOE 72] HOENEISEN B., MEAD C.A., "Limitations in microelectronics II. Bipolar technology", *Solid-State Electron*, vol. 15, pp. 819–829, 1972.

[JEA 89] JEANGUILLAUME C., COLLIEX C., "Spectrum-image: the next step in EELS digital acquisition and processing", *Ultramicroscopy*, vol. 28, 1989.

[KEL 07] KELLY T.K., MILLER M.K., "Invited review article: atom probe tomography", *Review of Scientific Instruments*, vol. 78, p. 031101, 2007.

[KOT 06] KOTULA P.G., KEEMAN M.R., "Application of multivariate statistical analysis to STEM X ray spectral images: interfacial analysis in microelectronics", *Microscopy and Microanalysis*, vol. 12, pp. 538–544, 2006.

[KRI 08] KRIVANEK O.L., CORBIN G.J., DELLBY N., ELSTON B.F., KEYSE R.J., MURFITT M.F., OWN C.S., SZILAGYI Z.S., WOODRUFF J.W., "An electron microscope for the aberration-corrected era", *Ultramicroscopy*, vol. 108, p. 179, 2008.

[LIG 09] LIGOR O., GAUTIER B., DESCAMPS A., ALBERTINI D., "Characterization of ultra-thin dielectrics using scanning capacitance microscopy", *AIP 2009 Conference: Frontiers of Characterization and Metrology for Nanoelectronics*, American Institute of Physics (ed), 2009.

[MAC 93] MACAULAY J.M., HULL R., JALALI B., "Characterization of arsenic doping profile across the polycrystalline Si/Si interface in polycrystalline Si emitter bipolar transistors", *Applied Physics Letters*, vol. 63, pp. 1258–1260, 1993.

[MUL 08] MULLER D.A., FITTING KOURKOUTIS L., MURFITT M., SONG J.H., HWANG H.Y., SILCOX J., DELLBY N., KRIVANEK O.L., "Atomic-scale chemical imaging of composition and bonding by aberration-corrected microscopy," *Science*, vol. 319, 2008.

[PAC 00] PACKAN P., AUDART N., "Toujours plus petit le transistor", *La Recherche*, no. 327, pp. 20–22, January 2000.

[PAN 11] PANTEL R., "Coherent bremsstrahlung effect observed during STEM analysis of dopant distribution in silicon devices using large area silicon drift EDX detectors and high brightness electron source", *Ultramicroscopy*, vol. 111, no. 11, pp. 1607–1618, 2011.

[PAN 12] PANTEL R., SERVANTON G., DELILLE D., KLENOV D.O., "High brightness electron source and large collection angle EDX SDD Detectors: A solution for fast elemental traces analysis and dopant mapping in silicon devices using the new Osiris TEM," *Proceedings from the 17th International Microscopy Congress (IMC17)*, Rio de Janeiro, September 19–25, 2012.

[PAR 08] PARISINI A., MORANDI V., SOLMI S., MERLI P.G., GIUBERTONI D., BERSANI M., VAN DEN BERG J.A., "Quantitative determination of the dopant distribution in Si ultrashallow junctions by tilted sample annular dark field scanning transmission electron microscopy", *Applied Physics Letters*, vol. 92, p. 261907, 2008.

[PEI 08] PEI L., DUSCHER G., STEEN C., PICHLER P., RYSSEL H., NAPOLITANI E., DE SALVADOR D., MARIA PIRO A.M., TERRASI A., SEVERAC F., CRISTIANO F., RAVICHANDRAN K., GUPTA N., WINDL W., "Detailed arsenic concentration profiles at Si/SiO2 interfaces", *Journal of Applied Physics*, vol. 104, p. 043507, 2008.

[RAP 55] RAPPAPORT P., LOFERSKI J.J., "Thresholds for electron bombardment induced lattice displacements in Si and Ge", *Physics Review*, vol. 100, p. 1261, 1955.

[RAU 99] RAU W.D., SCHWANDER P., BAUMANN F.H., HÖPPNER W., OURMAZD A., "Two-dimensional mapping of the electrostatic potential in transistors by electron holography", *Physics Review Letters*, vol. 82, pp. 2614–2617, 1999.

[SCH 10] SCHLOSSMACHER P., KLENOV D.O., FREITAG B., VON HARRACH S., STEINBACH A., "Nanoscale chemical compositional analysis with an innovative S/TEM-EDX system", *Microscopy and Analysis*, vol. 130, p. S5, 2010.

[SER 09] SERVANTON G., PANTEL R., JUHEL M., BERTIN F., "Arsenic 2D quantitative mapping in nanometer silicon devices using STEM EELS-EDX spectroscopy", *Micron*, vol. 40, pp.543-551, 2009.

[SER 10a] SERVANTON G., PANTEL R., "Arsenic dopant mapping in State-of-the-art semiconductor devices using electron energy loss spectroscopy", *Micron*, vol. 41, pp. 118–122, 2010.

[SER 10b] SERVANTON G., PANTEL R., JUHEL M., BERTIN F., "STEM EELS & EDX applications for quantitative arsenic dopant mapping in nanometer scale silicon devices," *Proceedings of the 16th Conference Microscopy of Semiconducting Materials,* March 2009, Journal of Physics: Conference Series, vol. 209, p. 012044, 2010.

[SER 12] SERVANTON G., PANTEL R., BERTIN F., "Arsenic dopant mapping in nanometer scale silicon devices using electron energy loss spectroscopy," *Proceedings from the 17th International Microscopy Congress (IMC17)*, Rio de Janeiro, September 19-25, 2012.

[STR 99] STRUDER L., MEIDINGER N., STROTTER D., KEMMER J., LECHNER P., LEUTENEGGER P., SOLTAU H., EGGERT F., ROHDE M., SCHULEIN T., "High resolution X-Ray spectroscopy close to room temperature", *Microscopy and Microanalysis*, vol. 4, pp. 622–631, 1999.

[SUN 05] SUNG C.Y., YIN H., NG H.Y., SAENGER K.L., CHAN V., CROWDER S.W., LI J., OTT J.A., BENDERNAGEL R., KEMPISTY J.J., KU V., LEE H.K., LUO Z., MADAN A., MO R.T., NGUYEN P.Y., PFEIFFER G., RACCIOPPO M., ROVEDO N., SADANA D., DE SOUZA J.P., ZHANG R., REN Z., WANN C.H., "High performance CMOS bulk technology using direct silicon bond (DSB) mixed crystal orientation substrates", *IEEE International, Electron Devices Meeting*, IEDM Technical Digest, 2005.

[THO 98] THOMPSON S., PACKAN P., BORH M., "Intel's 0.25 micron, 2.0 volts logic process technology", *Intel Technology Journal*, vol. Q3, p. 1, 1998.

[THO 05] THOMPSON K., BOOSKE J.H., BOOSKE J.H., LARSON D.J., KELLY T.F., "Three-dimensional atom mapping of dopants in Si nanostructures," *Applied Physics Letters*, vol. 87, p. 052108, 2005

[THO 06] THOMPSON K., LARSON D., BUNTON J., ULFIG R., PROSA T., SEBASTIAN J., OLSON J., LENS D., KELLY T., "Advanced analysis of Si nanostructures meet", *Abstr. Electrochem. Soc.*, vol. 601, 2006.

[THO 07] THOMPSON K., FLAITZ P.L., RONSHEIN P., LARSON D. J., KELLY T.F., "Imaging of arsenic Cottrell atmospheres around silicon defects by three-dimensional atom probe tomography", *Science*, vol. 317, no. 5843, pp. 1370–1374, 2007.

[THO 10] THOMAS P.J., TREVOR C., STROHBEHN E., WILBRINK J., TWESTEN R.D., "A Gubbens high-speed, hardware-synchronized STEM EELS spectrum-imaging using a next generation post-column imaging filter", *Microscopy and Microanalysis* , vol. 16, pp. 126–127, 2010.

[TIT 96] TITCHMARSH J., DUMBILL S., "Multivariate statistical analysis of FEG-STEM EDX spectra", *Journal of Microscopy*, vol. 184, pp. 195–207, 1996.

[TIT 99] TITCHMARSH J.M., "EDX spectrum modeling and multivariate analysis of sub-nanometer segregation", *Micron*, vol. 30, pp. 159–171, 1999.

[TOP 03] TOPURIA T., BROWNING N.D., MA Z., "Characterization of ultrathin dopant segregation layers in nanoscale metal-oxide-semiconductor field effect transistors using scanning transmission electron microscopy", *Applied Physics Letters*, vol. 83, pp. 4432–4434, 2003.

[TWI 03] TWITCHETT A.C., DUNIN-BORKOWSKI R.E., HALLIFAX R.J., BROOM R.F., MIDGLEY P.A., *Journal of Microscopy*, vol. 214, 2003

[VEL 06] VELLA A., VURPILLOT F., GAULT B., MENAND A., DECONIHOUT B., *Physics Review B*, vol. 73, p. 165416, 2006.

[VON 09] VON HARRACH H.S., DONA P., FREITAG B., SOLTAU H., NICULAE A., ROHDE M., *Microscopy and Microanalysis*, vol. 15, no. S2, 208, 2009.

[WAN 01] WANG T.S., CULLIS A.G., COLLART E.J.H, MURRELL A.J., FOAD M.A., *Microsc. Semicond. Mater. Conf.*, 2001.

[WIL 89] WILSON R.G., STEVIE F.A., MAGEE C.W., *Secondary Ion Mass Spectrometry: a Practical Handbook for Depth Profiling and Bulk Impurity Analysis*, Wiley Interscience, 1989.

[YAN 06] YANG M., CHAN V.W.C., CHAN K.K., SHI L., FRIED D.M., STATHIS J.H., CHOU A.I., GUSEV E., OTT J.A., BURNS L.E., FISCHETTI M.V., IEONG M., *IEEE transactions on electrons devices*, vol. 53, p. 5, 2006.

[ZHA 10] ZHANG L., SAITOH M., KOIKE M., TAKENO S., TANIMOTO H., ADACHI K., YASUTAKE N., KUSUNOKI N., *Extended Abstracts of 2010 International Workshop on Junction Technology*, IEEE press, pp. 1–5, 2010.

Chapter 3

Quantitative Strain Measurement in Advanced Devices: A Comparison Between Convergent Beam Electron Diffraction and Nanobeam Diffraction

3.1. Introduction

To enhance carrier mobility and drive current in metal-oxide–semiconductor field effect transistors, several approaches, mainly based on strain engineering have been used for several years [GHA 03, ANT 06, LEE 05, THO 06, PAY 08]. Among these, selective epitaxial growth of raised boron doped silicon–germanium in the transistor source and drain regions is proven to induce a longitudinal compressive stress in the Si channel [SMI 05, YEO 05]. A great challenge in the process development of advanced microelectronic devices is now to assess the in situ stress state in the channel region and to link the obtained results with electrical tests. With high resolution electron microscopy [HŸT 98, HUE 08] and more recently dark-field holography technique [HŸT 08], convergent beam electron diffraction (CBED) [ARM 03] and nanobeam diffraction (NBD) [USU 04] in a transmission electron microscope (TEM) have emerged as two of the best ways to measure strain at a nanometer scale. In this chapter, these two diffraction-based techniques are used to study the strain induced by the SiGe source and drain in sub-45 nm devices. Spatial resolution and sensitivity capabilities will be discussed. The main scope of this chapter is to compare first strain measurement provided by both techniques (CBED

Chapter written by Laurent CLEMENT and Dominique DELILLE.

and NBD) and second, to correlate strain results with electrical tests performed on the same devices. The effect of gate length on stress state in the Si channel is discussed and a comparison between numerical solutions and experimental measurements is performed.

3.2. Electron diffraction technique in TEM (CBED and NBD)

CBED has been widely used for several years to investigate stress and strain fields in microelectronic devices [ZHA 06, LI 06]. Considered for a long time as one of the most powerful techniques to measure lattice parameters, the CBED technique suffers from some drawbacks which we describe below.

Although less sensitive to lattice parameters variation, NBD offers some advantages compared to CBED and recent works have shown very interesting results [LIU 08, BEC 09].

In this section, these two techniques applied to strain measurement are successively described. The focus is on the different tools that have been developed for CBED and NBD patterns post-treatment, to accurately determine strain field in the investigated materials.

3.2.1. *CBED patterns acquisition and analysis*

Strain measurement by the CBED technique consists in analyzing the position of the high order Laue zone (HOLZ) lines, which mainly depends on the crystal lattice parameters. The sample has to be tilted from dynamical zone axis by several degrees ($\sim 10°$) and an electron beam with a semi-convergence angle of about 5–15 mrad is focused on the sample using a 50–100 μm C_2 aperture in a two-condenser system (Figure 3.1). The obtained probe (around 1-nm size) allows acquiring CBED patterns series on the charge-coupled device (CCD) camera using scanning TEM mode. The sample can also be cooled down during the experiments to get sharper HOLZ lines, high S/N ratio and a better contrast in CBED patterns by lowering the Debye–Waller factor (Figure 3.2(b)). With Hough's algorithm, HOLZ lines can be extracted from the full CBED patterns, and by numerically comparing the experimental HOLZ lines positions with simulated ones, strain field in the observed crystal can be determined with a sensitivity of 2×10^{-4} [SEN 03, ZUO 92].

Nevertheless, the main drawback of this technique (and in fact of all TEM-based techniques) is the necessity of thin lamella preparation that induces stress relaxation at the created free surfaces. Therefore, CBED patterns can be disturbed due to a non-uniform strain through the specimen thickness. Varying Bragg condition along the electron beam induces HOLZ lines broadening and the treatment of such patterns becomes more complex [CLE 04, HOU 06]. We have already reported some works on this subject and we have shown that by combining finite element modeling of stress relaxation in TEM lamella with dynamical electron diffraction, HOLZ lines broadening can be reproduced. We have then proposed an original procedure for strain measurement in complex devices through the quantification of this stress relaxation in a TEM sample [CLE 09].

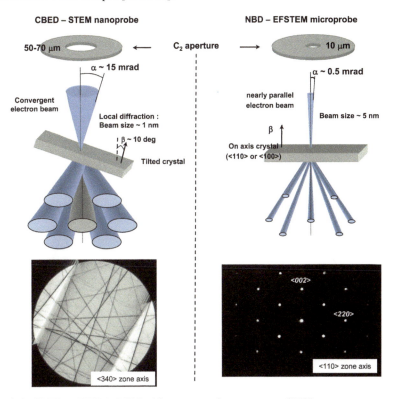

Figure 3.1. *CBED vs NBD in TEM with a two-condenser system. CBED patterns are acquired in the STEM nanoprobe mode with a large semi-convergence angle ($\alpha = 5\text{--}15$ mrad) and a 1-nm probe size. Only the transmitted disk with fine HOLZ lines is studied. NBD patterns are acquired in the EFSTEM microprobe mode with a nearly parallel electron beam ($\alpha = 0.5$ mrad) and a 5-nm probe size*

In this chapter, only CBED patterns with fine HOLZ lines are analyzed. As a matter of fact, in some cases, due to device geometry symmetry, such patterns can be obtained. Classical approach for strain measurement can then be used. It consists in (1) the acquisition of CBED patterns with fine HOLZ lines, (2) the numerical extraction of experimental HOLZ lines by using the Hough transform algorithm, (3) the quasi-kinematical simulation of HOLZ lines position, (4) the minimization on lattice parameters of an error function defined between experimental and simulated HOLZ lines position and, (5) finally, to confirm result from step (iv), dynamical simulation of the whole CBED pattern can be performed using found lattice parameters to further enhance the accuracy of the determination of the HOLZ line shifts and, thus, of the quantification of the strain in the device. The two different tools (based on both kinematical and dynamical simulations) have been developed to analyze CBED patterns under the Matlab®environment as described below.

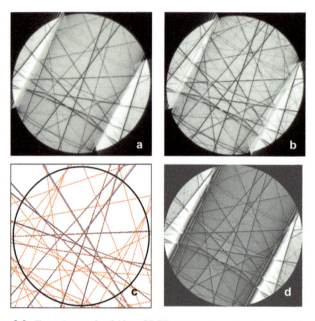

Figure 3.2. *Experimental <340> CBED patterns acquired in Si without and with cooling down the sample (a and b, respectively). Corresponding kinematical and dynamical simulations using Bloch waves formalism (c and d, respectively)*

3.2.1.1. Strain determination using quasi-kinematical simulation of HOLZ lines and Hough transform

In the kinematic approximation, the HOLZ line position is given by the Bragg law as [ZUO 92]:

$$2\mathbf{K} \cdot \mathbf{S_g} = \mathbf{K}^2 - (\mathbf{K} + \mathbf{g})^2 = 0 \qquad [3.1]$$

K is the mean wave vector inside the crystal, **g** is a reciprocal lattice vector and **S_g** is the excitation error equal to zero at zero-order Laue zone (HOLZ) line positions. In a (x, y, z) orthogonal coordinate system with x-axis as one of the **h** vector, in the ZOLZ and z-axis as the direction of the nearest zone axis, the equation of the HOLZ line given by equation [3.1] is:

$$K_x g_x + K_y g_y = K g_z - \frac{g^2}{2} \qquad [3.2]$$

Thus, the HOLZ line position is a function of crystal lattice parameters and accelerating voltage through, respectively, **g** components and **K** vector. HOLZ lines for a particular zone axis can then be easily plotted and superposed to the ones extracted from experimental CBED patterns using Hough transform [DUD 72] as shown in Figure 3.2(c). A minimization algorithm on lattice parameters between experimental and simulated HOLZ lines intersection points is then applied to match both line positions assuming some hypothesis that we will discuss in section 3.4.1.1.

In practice, we generally need two steps for such analysis: (1) first, the effective accelerating voltage E_{eff} has to be determined by analyzing a CBED pattern acquired in a no-strained silicon region with well-known lattice parameters. In this case, the minimisation between the extracted and simulated HOLZ lines position is performed on the accelerating voltage E. (2) Assuming this value keep constant, HOLZ line shift in CBED patterns acquired in the different strained regions can be simulated by fitting line positions on lattice parameters. A local strain field can thus be deduced from this well-known procedure. Details on the different algorithms used (Hough transform, minimization), out of the scope of this chapter, are not discussed here but can be found elsewhere [CLE 06].

3.2.1.2. *Strain measurement refinement using dynamical simulation of CBED patterns including absorption and temperature effect*

In the Bloch wave approach, for a given incident wave $\boldsymbol{K}_0$, the wave function $\Psi(\boldsymbol{r})$, solution of the Schrödinger equation for a periodic crystal, can be decomposed into a linear superposition of three-dimensional (3D) Bloch waves (j) that are characterized in the plane wave basis $\boldsymbol{g}$ by the coefficients $C_{\boldsymbol{g}}^{(j)}$ and $\gamma^{(j)}$:

$$\Psi(\boldsymbol{r}) = \sum_j \epsilon^{(j)} \sum_{\boldsymbol{g}} C_{\boldsymbol{g}}^{(j)} \exp(2\pi i (\boldsymbol{K} + \gamma^{(j)}\boldsymbol{n} + \boldsymbol{r})\boldsymbol{g} \qquad [3.3]$$

where $\boldsymbol{n}$ is the outward normal to the entrance surface and $\boldsymbol{K}$ is the incident wave vector inside the crystal, that the incident wave $\boldsymbol{K}_0$ corrected by the mean inner potential U_0 ($|\boldsymbol{K}|^2 = |\boldsymbol{K}_0|^2 + U_0$ with the equal components of $\boldsymbol{K}$ and $\boldsymbol{K}_0$ in the entrance plane).

Using this expression in the Schrödinger equation leads to the electrons dispersion equation inside the crystal [SPE 92]:

$$2\mathbf{K} \cdot \mathbf{S_g} C_g^{(j)} + \sum_{h \neq g} U_{g-h} C_h^{(j)} = 2\left[(\mathbf{K} + \mathbf{g}) \cdot \mathbf{n}\right] \gamma^{(j)} C_g^{(j)} \qquad [3.4]$$

From a practical point of view, a limited number of N beams g is generally used in the calculations therefore the coefficients $C_g^{(j)}$ then define a N-column vector $C^{(j)}$ in a matrix representation. $C^{(j)}$ and $\gamma^{(j)}$ are, respectively, eigenvectors and eigenvalues of the secular equation:

$$\mathbf{A}\mathbf{C}^j = \mathbf{B}\mathbf{I}\gamma^{(j)} \mathbf{C}^j \qquad [3.5]$$

where the $N \times N$ **A** and **B** matrix are defined as:

$$\mathbf{A}_{gh} = 2\mathbf{K} \cdot \mathbf{S_g} \delta_{gh} + U_{g-h}(1 - \delta_{gh}) \qquad [3.6]$$

$$\mathbf{B}_{gh} = 2\left[(\mathbf{K} + \mathbf{g}) \cdot \mathbf{n}\right] \delta_{gh} \qquad [3.7]$$

U_g are proportional to Fourier transform components V_g calculated for each reflection **g** from atomic scattering factor expression $f^{(B)}$ and Debye-Waller value M as:

$$V_g = \frac{h^2}{2\pi m_0 e \Omega} \sum_j f_j^B(\mathbf{s}) \exp(-M_j \mathbf{s}^2) \exp(-2i\pi \mathbf{g} \cdot \mathbf{r}_j) \qquad [3.8]$$

where the sum is over the atoms inside the lattice. Weickenmeier and Kohl [WEI 91] have proposed an analytical expression of the atomic scattering factors $f^{(B)}$ that we have implemented in our code. The Debye-Waller values used here are, respectively, $M_{Si} = 0.49$ Å^2 at room temperature (300 K) and $M_{Si} = 0.2$ Å^2 at 100 K [PEN 96]. Figure 3.2(d) shows an example of a whole CBED pattern dynamical simulation including absorption and temperature effects and with approximately 90 beams in the calculation. More details on the dynamical simulation of CBED patterns can be found elsewhere [CLE 06].

3.2.2. *NBD patterns acquisition and analysis*

The NBD technique uses a nearly parallel electron beam that can be obtained either by using a three-condenser system [SOU 09] or by lowering the C_2 aperture size down to about 10 μm (Figure 3.1). Armigliato *et al.* show that a 1-μm C_2 aperture can even be used to reduce a convergence angle to 0.1 mrad [ARM 08]. Diffraction spots are thus formed at the back-focal plane of the objective lens by focusing the beam on the specimen, and can be acquired in the STEM mode. The NBD technique allows a direct sample strain measurement using the shifts in diffraction spots as a function of the beam position on the sample and with respect to a reference diffraction pattern. No direct knowledge of the sample remains necessary and any zone axis can be selected on relatively thick samples although sample thickness will be limited mainly by the signal strength and instrument stability. An energy filter can also be used to lower the contribution of inelastically scattered electrons. The main limitations in the use of the NBD technique are the following: first, contrary to CBED where HOLZ line shifts are measured, the NBD technique cannot determine the 3D strain field since only ZOLZ reflections are analyzed in NBD patterns. Nevertheless, recording diffraction patterns in the direction (mainly <110> or <100> zone axis) along which microelectronic devices are oriented does not degrade spatial resolution that only depends in this case on the probe size of about 5 nm. Second, the sensitivity of the NBD technique remains poor when compared to CBED since a typical TEM diffraction pattern collected on a 1,024 × 1,024 pixels CCD camera will typically require subpixel accuracy to measure strain of less than 0.2%. In the present study, by utilizing the full diffracted pattern and careful selection of a peak finding algorithm, the software package TrueCrystal Strain (FEI Company) has demonstrated strain resolution higher than 0.2% [SOU 09].

3.3. Experimental details

3.3.1. *Instrumentation and setup*

The present studies were performed on a FEI TECNAI F20 TEM operating at a high voltage of 200 kV and equipped with a Gatan Imaging Filter (GIF 2000). TEM samples were extracted from electrically tested integrated circuit (ICs) and thinned down to about 300–400 nm using the in situ lift-out technique with an FEI Helios focused ion beam (FIB) at 30 kV.

CBED patterns have been acquired in STEM nanoprobe mode along the <340> zone axis tilted by 8.4° from the <110> cross-section direction. The sample was cooled down during the experiments for better post-treatment efficiency as explained in section 3.2.1.

During NBD experiments, the sample was oriented along the <110> direction and the STEM microprobe mode was used. By inserting a 10-μm C_2 condenser aperture, a nearly parallel electron beam (0.3 mrad semi-convergence angle) was obtained with a probe size of about 5 nm.

For both CBED and NBD pattern acquisitions, the contribution of inelastically scattered electrons was lowered using an imaging filter.

3.3.2. Samples description

A first study was conducted on a fully processed pMOS transistor with a classical 40-nm poly-silicon gate length and integrating the $Si_{0.8}Ge_{0.2}$ source and drain regions. For this device named *device 1*, both CBED and NBD experiments were performed to evaluate strain state in the Si channel region. Experimental results will be presented in section 3.4 and a direct comparison of the two techniques on the same sample will be discussed.

The second sample named *device 2* is composed of SiGe stressors and a totally nickel silicide gate. The interest of using such a metal gate is to limit the poly-Silicon depletion phenomenon occurring with a classical gate stack ($Si/SiO_2/poly$-Si). NBD experiments have been carried out on this sample and strain components through the Si channel were measured as a function of gate length. A correlation between strain measurement and mobility extracted from electrical test is thus found.

3.4. Results and discussion

3.4.1. *Strain evaluation in a pMOS transistor integrating eSiGe source and drain – a comparison of CBED and NBD techniques*

3.4.1.1. *CBED experiments and results*

Figure 3.3(a) shows a STEM image of the transistor with the different points labeled 1–7 where CBED patterns have been acquired. Two experimental diagrams acquired at positions 1 and 6 and corresponding, respectively, to the reference region without strain far from the transistor gate and the region between source and drain in the Si channel are shown in Figure 3.3(b). A comparison between the two diagrams clearly shows HOLZ lines shift as underlined by the shape difference of the two triangles drawn at the center of both CBED patterns. Quasi-kinematical simulation was first performed to determine Si lattice parameters variation between these two regions assuming a compressive stress state along the <110> as expected in this process configuration. This geometry symmetry leads to some relations between lattice parameters variation: $\Delta a = \Delta b$ and $\alpha = \beta = \pi/2$. These relations reduce to three, the number of independent parameters, to fit in the simulation. Note that the value of γ could be also deduced from geometrical consideration and should lead to the additional relation: $\Delta\gamma/2 = \Delta a/a$. However, we have preferred to keep the γ value as a fitting parameter in our simulation for a better accuracy of HOLZ lines shift.

Once the lattice parameters are extracted from the minimization algorithm, a dynamical simulation of the full CBED pattern is systematically performed with the fitting values previously obtained to confirm the determination of the HOLZ lines shift. These simulations are shown in Figure 3.3(b) (on the right) and compared with experimental CBED patterns 1 and 6. HOLZ lines shift at the center of each diagram is well reproduced by the simulation indicating the strain state in the corresponding Si region can be determined very accurately.

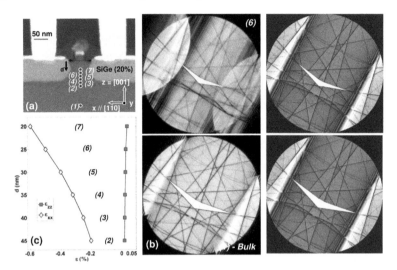

Figure 3.3. *CBED results on poly-Si gate pMOS transistor with eSiGe (device 1). a) STEM image of the investigated device. b) Experimental <340> CBED patterns acquired, respectively, at positions (1) and (6) and corresponding dynamical simulations. c) Strain component values along x and z directions as a function of depth d under the gate calculated from HOLZ line shifts*

The graph in Figure 3.3(c) shows the results deduced from all the CBED patterns analysis in terms of strain components along the x (<110>) direction parallel to gate length and z (<001>) direction. A compressive strain state along x direction: $\epsilon_{xx} < 0$ (and the elastic corresponding tensile strain along z) is clearly measured in the Si channel as expected. This compressive strain ϵ_{xx} component increases when approaching the poly-Si gate and reaches -0.6% at $d = 30$ nm from the gate. The corresponding stress value is then around $\sigma_{xx} = -900$ MPa.

Unfortunately, no CBED patterns with fine and well-defined HOLZ lines could be acquired very close to the Si/gate interface. Moreover, the very small number of "well suited" CBED patterns for HOLZ line shift measurement (8 in this line-scan) have been obtained exactly in the middle of the Si channel (at equal distance from the SiGe source and drain). For any other positions, HOLZ lines broadening occurs in the diagrams. These two remarks raise the main drawback of CBED experiments: most of experimental patterns acquired in such complex devices with a lot of different material interfaces are too blurred (due to HOLZ lines broadening) to be correctly studied. In fact, HOLZ lines are very sensitive to atomic plane bending, which occurs at the free interfaces of the TEM lamella. Even if the resultant HOLZ line broadening can be simulated by coupling dynamical simulation with finite element modeling, the analysis then becomes more complex and strain quantification in this case remains very difficult. Although very sensitive to lattice parameters change (in the case of fine HOLZ lines) – a variation of strain component $\epsilon_{zz} < 0.05\%$ can be easily measured (Figure 3.3(c)) – HOLZ lines broadening strongly limit CBED as a straightforward technique for quantitative strain measurement in real devices. So, in this case, the NBD technique can be an alternative to CBED.

3.4.1.2. *NBD experiments and results*

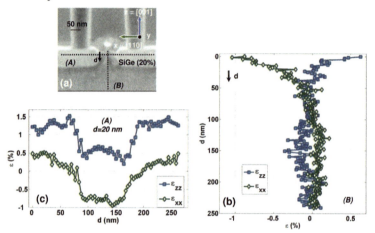

Figure 3.4. *NBD results on poly-Si gate pMOS transistor with eSiGe (device 1). a) STEM μ–probe image of the investigated device. Strain component values along x and z directions along line scan (B) under the gate b) and along line scan (A) through source, channel and drain regions c)*

Figure 3.4 shows results obtained using the NBD technique on the same device analyzed previously by CBED. A STEM image of the structure is also presented in Figure 3.4(a) but here in μ-probe mode used for NBD experiments that explains the poor resolution. Two lines scan (A) and (B), respectively, parallel and perpendicular to the device have been recorded with about 200 NBD patterns for both. TrueCrystal Strain software has been used to determine directly for each measured point the two main strain components ϵ_{xx} and ϵ_{zz} as plotted in Figures 3.4(b) and 3.4(c) like explained in section 3.2.2. This shows the first advantage of NBD compared to CBED where the strain field components along the main device axis cannot be measured directly since only lattice parameters variation in the crystal axis system can be obtained. Moreover, contrary to CBED results where only few points are measured, NBD experiments post-treatment give us more information in regions spreading from the substrate to the Si/poly-Si interface (line scan (B)) and through the whole channel length and even in the SiGe source and drain regions (line scan (A)). Although more noisy than CBED results, the compressive ϵ_{xx} strain component is well measured and reaches -1% at the poly-Si gate/Si interface corresponding to a stress value of $\sigma_{xx} = -1.2$ GPa.

3.4.2. *Quantitative strain measurement in advanced devices by NBD*

3.4.2.1. *Application to a 45-nm metal gate pMOS transistor with eSiGe*

Figure 3.5 shows a STEM HAADF image of *device 2* and results obtained using the NBD technique. Horizontal line scan (A) with the two main strain components (ϵ_{xx} and ϵ_{zz}) through the Si channel length and the source and drain regions at a distance $d = 20$ nm from the bottom of the metal gate is presented. Two other scans (B) and (C) in the perpendicular direction and located, respectively, at the middle of the gate and below one stressor are also shown. Once again, a compressive strain along the x direction in the Si channel induced by the stressors is clearly evidenced by these experiments. The study of line scan (C) also shows the sign inversion of strain (compressive along z and tensile along x) in the silicon region just below the SiGe stressor. This result is coherent with finite element simulation conducted elsewhere [YEO 05].

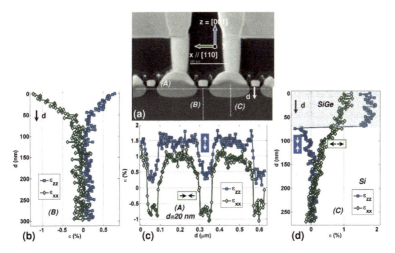

Figure 3.5. *NBD results on a 45-nm metal gate pMOS transistor with eSiGe (device 2). a) STEM image of the investigated device. Strain component values along x and z directions along line scan (B) under the gate b), along line scan (A) through source, channel and drain regions (b) and along line scan (C) just below one stressor d)*

3.4.2.2. Influence of gate length

To better understand stressor integration in advanced devices and to correlate experimental results with mechanical simulations, the strain induced by the SiGe stressors as a function of gate length has been studied too. Figure 3.6 shows ϵ_{xx} and ϵ_{zz} strain components profile through the Si channel (horizontal scan) for two gate lengths (220 and 480 nm, respectively). Contrary to results obtained on the 45-nm gate length transistor where ϵ_{xx} and ϵ_{zz} are quietly constant through the Si channel, the absolute values of strain components are maximum very close to the SiGe source and drain regions and then decrease when approaching middle of the gate. Far away from stressor regions, the residual strain measured ($\epsilon_{xx} = -0.05\%$ and $\epsilon_{zz} = 0.1\%$) at the middle of the Si channel for the larger transistor is probably due to the metal gate and not anymore by the SiGe source and drain.

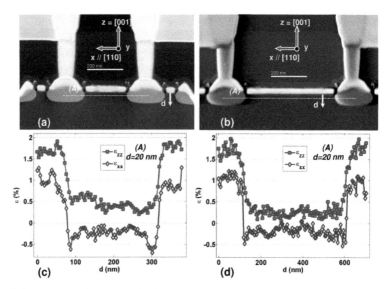

Figure 3.6. *Influence of gate length (device 2). STEM images of the two transistors with gate length 220 and 480 nm (a and b, respectively). Strain component values along x and z directions along line scan (A) through source, channel and drain regions for the two transistors (c and d, respectively)*

3.4.2.3. Correlation between mechanical simulation and electrical measurement

Figure 3.7(a) shows a superposition of experimental strain results obtained on the three different transistor gate lengths of the investigated *device 2*. Mechanical simulation using FEM in COMSOL Multiphysics software [COM 08] (Figure 3.7(b)) is performed. The numerical solution is compared with experimental measurements. Qualitative agreement between experiment and simulation is rather good. However, mechanical calculations seem to overestimate strain value in the Si channel compared to experimental results. Nevertheless, a perfect matching between simulation and experiments is probably difficult to achieve because of the lack of mechanical knowledge of the specimen (intrinsic stress in the gate, contribution of the free interfaces for the 3D specimen relaxation).

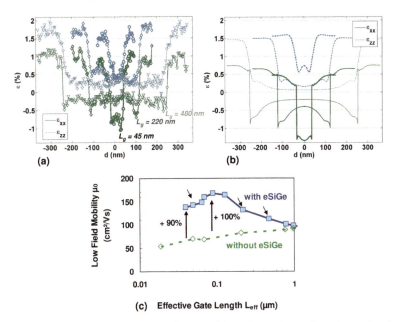

Figure 3.7. *Experience vs simulation. a) Superposition of experimental strain results obtained on the three different transistor gate lengths of the investigated device 2. b) Strain components profile extracted from finite element method (FEM) of this structure c) Mobility measurement as a function of gate length, with and without the SiGe stressor process integration. The three arrows in this graph correspond to the three different gate lengths investigated here by the NBD technique*

Figure 3.7(c) finally shows mobility properties of such advanced devices as a function of gate length, with and without the SiGe stressor process integration. The three arrows in this graph correspond to the three different gate lengths investigated here by the NBD technique. The influence of SiGe source and drain regions integration in terms of mobility enhancement is clearly evidenced here for gate length $Lg < 1$ μm as expected. The slight mobility decrease when decreasing gate length from 100 nm is due to the short channel effect (SCE) phenomenon that we will not discuss here. Despite this, mobility gain is still relevant for small transistor gate lengths when compared with the classical process without stressor.

3.5. Conclusion

In this chapter, we compared CBED and NBD techniques applied for strain determination on a fully processed pMOS transistor integrating the SiGe source and drain regions. By cooling down the sample and using dynamical simulations for fine refinement, Si lattice parameters in the transistor's channel can then be measured accurately (with a sensitivity of 2×10^{-4}) by CBED. Nevertheless, we also show that such measurements are quite difficult to obtain since HOLZ lines broadening usually occurs in CBED patterns due to atomic planes bending along the electron beam. Then, a direct measurement of lattice parameters and therefore strain field is not possible anymore unless by coupling experiments with mechanical simulations as already mentioned. We then show that, in this case, the NBD technique can be an alternative to CBED. By using TrueCrystal Strain software from FEI, we demonstrate that a direct measurement of the main interested strain components can be obtained. No sample tilt is required anymore and no complex dynamical simulation has to be performed to measure strain value with a sensitivity lower than 0.2%.

To better understand stressors' integration in the next-generation MOS process, the NBD technique has been used to correlate strain field with mobility measurement as a function of gate length. Mechanical simulations have also been performed by the finite element method of the real 3D transistor geometry results are in good agreement with experiments. NBD is therefore a well-suited technique for strain field determination in "real" devices with high stress level process integration. For a low stressed buried layer and for less complex structure geometry, the CBED technique remains nevertheless a powerful technique for strain field determination with a better sensitivity than NBD.

3.6. Bibliography

[ANT 06] ANTONIADIS D.A., *IBM Journal of Research and Development*, vol. 50, p. 363, 2006.

[ARM 03] ARMIGLIATO A., BALBONI R., CARNEVALE G.P., PAVIA G., PICCOLO D., FRABBONI S., BENEDETTI A., CULLIS A.G., *Applied Physics Letters*, vol. 82, p. 2172, 2003.

[ARM 08] ARMIGLIATO A., FRABBONI S., GAZZADI G., *Applied Physics Letters*, vol. 93, p. 161906, 2008.

[BEC 09] BECHE A., ROUVIERE J.L., CLEMENT L., HARTMANN J.M., *Applied Physics Letters*, vol. 95, p. 123114, 2009.

[CLE 04] CLEMENT L., PANTEL R., KWAKMAN L.F.TZ, ROUVIERE J.L., *Applied Physics Letters*, vol. 85, p. 651, 2004.

[CLE 06] CLEMENT L., PhD Thesis, 2006.

[CLE 09] CLEMENT L., CACHO F., PANTEL R., ROUVIERE J.-L., *Micron*, vol. 40, p. 886, 2009.

[COM 08] COMSOL MULTIPHYSICS, 2008, available at http://www.comsol.com.

[DUD 72] DUDA R., HART P., *Communication of the ACM*, vol. 15, p. 11, 1972.

[GHA 03] GHANI T., *IEDM Technical Digest*, p. 978, 2003.

[HŸT 98] HŸTCH M., SNOECK E., KILAAS R., *Ultramicroscopy*, vol. 74, p. 131, 1998.

[HŸT 08] HŸTCH M., HOUDELLIER F., SNOECK E., *Nature*, vol. 453, p. 1086, 2008.

[HOU 06] HOUDELLIER F., ROUCAU C., CLEMENT L., ROUVIERE J.L., CASANOVE M.J., *Ultramicroscopy*, vol. 106, p. 951, 2006.

[HUE 08] HUE F., HŸTCH M., BENDER H., HOUDELLIER F., CLAVERIE A., *Physical Review Letters*, vol. 100, p. 156602, 2008.

[LEE 05] LEE M.L., FITZGERALD E.A., BULSARA M.T., CURRIE M.T., LOCHTEFELD A., *Journal of Applied Physics*, vol. 97, p. 011101, 2005.

[LI 06] LI J., DOMENICUCCI A., CHIDAMBARRAO D., GREENE B., ROVDEDO N., HOLT J., DUNN D., NG H., *Materials Research Society Symposium Proceedings*, vol. 913, 2006.

[LIU 08] LIU J.P., LI K., PANDEZ S.M., BENISTANT F.L., SEE A., ZHOU M.S., HSIA L.C., SCHAMPERS R., KLENOV D.O., *Applied Physics Letters*, vol. 93, p. 221912, 2008.

[PAY 08] PAYET F., BOEUF F., ORTOLLAND C., SKOTNICKI T., *IEEE Transactions on Electron Devices*, vol. 55, 2008.

[PEN 96] PENG L.-M., REN G., DUDAREV S.L., WHELAN M.J., *Acta Crystallographica*, vol. A52, p. 456, 1996.

[SEN 03] SENEZ V., ARMIGLIATO A., DEWOLF I., CARNEVALE C., BALBONI R., FRABBONI S., BENEDETTI A., *Journal of Applied Physics*, vol. 94, p. 5574, 2003.

[SMI 05] SMITH L., MOROZ V., ENEMAN G., VERHEYEN P., NOURI F., WASHINGTON L., JURCZAK M., PENSZIN O., PRAMANIK D., DE MEYER K., *IEEE Electron Device Letters*, vol. 26, p. 652, 2005.

[SOU 09] SOURTY E., STANLEY J., FREITAG B., *IEEE Proceedings of 16th IPFA*, p. 479, 2009.

[SPE 92] SPENCE J.C.H., ZUO J.M., *Electron Microdiffraction*, Plenum Press, New York, 1992.

[THO 06] THOMPSON S.E., SUN G.Y., CHOI Y.S., NISHIDA T., *IEEE Transactions on Electron Devices*, vol. 53, p. 1010, 2006.

[USU 04] USUDA K., MIZUNO T., TEZUKA T., SUGIYAMA N., MORIYAMA Y., NAKAHARAI S., TAKAGI S., *Applied Surface Science*, vol. 224, p. 113, 2004.

[WEI 91] WEICKENMEIER A., KOHL H., *Acta Crystallographica*, vol. A47, p. 590, 1991.

[YEO 05] YEO Y.-C., SUN J., *Applied Physics Letters*, vol. 93, p. 023103, 2005.

[ZHA 06] ZHANG P., ISTRATOV A.A., WEBER E.R., KISIELOWSKI C., HE H., NELSON C., SPENCE J.C.H., *Applied Physics Letters*, vol. 89, p. 161907, 2006.

[ZUO 92] ZUO J.M., *Ultramicroscopy*, vol. 41, p. 211, 1992.

Chapter 4

Dark-Field Electron Holography for Strain Mapping

4.1. Introduction

Strained silicon is now an integral feature of the latest generation of transistors and electronic devices [ITR 11, GHA 03, ANT 06] because of the associated enhancement in carrier mobility [LEE 05, THO 06]. Strain is also expected to have an important role in future devices, for example optoelectronic components [JAC 06]. Different strategies have been used to engineer strain in devices, leading to complex strain distributions in two and three dimensions [ACO 06, PAR 06]. Developing methods of strain measurement at the nanoscale has, therefore, been an important objective in recent years because, at the time, none of the existing techniques combined the necessary spatial resolution, precision and field of view [ITR 11, FOR 06]. For example, Raman spectroscopy or X-ray diffraction techniques can map strain at the micrometric scale, whereas transmission electron microscopy (TEM) allows strain measurement at the nanometer scale but only over small sample areas. The technique described in this chapter, dark-field electron holography (DFEH), was developed specifically to solve this problem and measure strain to high precision, with nanometer spatial resolution and for micrometer fields of view [HYT 08, HYT].

TEM is the only tool capable of measuring strain at the nanoscale; the techniques fall into two broad categories: diffraction-based and image-based. Convergent beam

---

Chapter written by Martin HŸTCH, Florent HOUDELLIER, Nikolay CHERKASHIN, Shay REBOH, Elsa JAVON, Patrick BENZO, Christophe GATEL, Etienne SNOECK and Alain CLAVERIE.

electron diffraction (CBED) is the oldest of these and has been applied to study strain in devices [ZHA 06]. However, CBED is indirect and relies on detailed comparisons between experimental data and simulation. More importantly, it often fails in the highly strained active regions of devices due to bending of the lattice planes [CLE 04, HOU 06]. Nanobeam electron diffraction (NBED) is a more recent technique and shows promise for more flexible characterization of structures, though remains less precise than CBED [USU 05]. Diffraction techniques are applied to produce isolated point-by-point measurements and are difficult to apply to map strain across large areas.

Of the imaging techniques, geometric phase analysis (GPA) of high-resolution transmission electron microscope (HRTEM) images has been very successful in mapping strain in nanometric-sized areas [HYT 98, HYT 03, JOH 08]. It can be extended to larger areas by reducing the magnification and correcting for the distortions of the projector lens that becomes important at lower magnification [HUE 05]. These algorithms and other developments have been incorporated into commercially available software, popularizing the use of the technique [GPA]. The strain in the active region of strained silicon devices was mapped for the first time in such a way [HUE 08]. However, the field of view is still limited with respect to the devices and areas we would like to characterize (see Figure 4.1(a)) and cannot be improved much further given the need to image the atomic lattice. This problem led to the development of DFEH (see Figure 4.1(b)) [HYT 08].

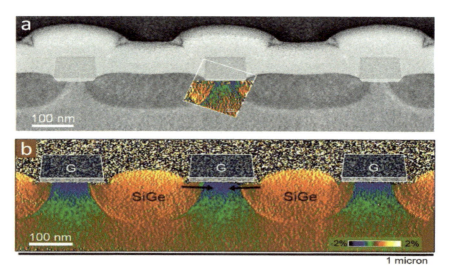

Figure 4.1. *The problem of the scale of strain measurements: a) strain map obtained by HRTEM (in color) superimposed on a bright-field image of the full structure [HUE 08]; b) strain map obtained by DFEH of a similar device [HYT 08]*

4.2. Setup for dark-field electron holography

The setup for DFEH is shown in Figure 4.2, along with the conventional setup for off-axis electron holography (CEH) [HYT 11]. A parallel beam is formed by the condenser system to impinge on the specimen, directly in the case of conventional holography, or at a slight angle in the case of dark-field holography. One part of the beam passes through a reference region, a vacuum in the case of conventional holography, or an unstrained region of crystal for dark-field holography. The other part travels through the region of interest. An electrostatic biprism is then used to interfere the two parts of the wave, the transmitted wave in the case of CEH, or the diffracted wave for DFEH. Indeed, apart from the biprism, the two setups are identical to a conventional bright-field and dark-field experiment, respectively. The interference fringes viewed at the screen encode the phase difference between the two wave paths.

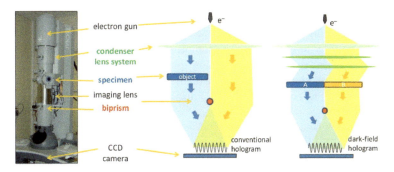

Figure 4.2. *Schematic representation of both conventional (center) and dark-field (right) off-axis electron holography setups. In the conventional setup, the reference is made with the vacuum whereas for dark-field holography, the reference is a region of unstrained crystal (crystal A). The diffracted beam from the reference is interfered with the diffracted beam from the region of interest (crystal B) to form the dark-field hologram.*

The phase has a number of contributions and can be written as a sum of four principal terms [HYT 11]:

$$\phi(\mathbf{r}) = \phi^G(\mathbf{r}) + \phi^C(\mathbf{r}) + \phi^M(\mathbf{r}) + \phi^E(\mathbf{r}) \qquad [4.1]$$

of which the crystalline phase, $\phi^C(\mathbf{r})$, encompassing dynamic diffraction phase factors, and phases from magnetic, $\phi^M(\mathbf{r})$, electrostatic fields $\phi^E(\mathbf{r})$ are well known. Geometric phase, $\phi_g^G(\mathbf{r})$, which arises from the presence of strain and displacement fields, is defined as:

$$\phi_g^G(\mathbf{r}) = -2\pi \mathbf{g}.\mathbf{u}(\mathbf{r}) \qquad [4.2]$$

where $\mathbf{g}$ is the diffraction vector and $\mathbf{u}(\mathbf{r})$ the displacement field as a function of position. Note that the transmitted beam, for which $\mathbf{g} = 0$, does not carry information about displacement fields. Only diffracted beams carry geometric phase, which lead to the idea of DFEH to measure the phase of diffracted beams in order to study strain.

It can be seen also from equation [4.2] that only displacements that are parallel to the diffraction vector are measured. This is fine if you want to measure strains in the diffracted beam direction, $\mathbf{g}$. The strain component giving the compression and tension in this direction, ε_{gg}, is given simply by:

$$\varepsilon_{gg} = \nabla_g u_g = \frac{-1}{2\pi g}\nabla_g \phi_g^G \qquad [4.3]$$

where u_g is the component of the displacement field in the direction of $\mathbf{g}$, and ∇_g is the gradient parallel to $\mathbf{g}$. A common example is to use the (004) diffracted beam to measure strains in the growth direction [001]. The full strain tensor in two dimensions requires, however, the measurement of the geometric phase from two diffracted beams. The equation for the 2D displacement field is as follows:

$$\mathbf{u}(\mathbf{r}) = -\frac{1}{2\pi}\left[\phi_{g1}^G(\mathbf{r})\mathbf{a}_1 + \phi_{g2}^G(\mathbf{r})\mathbf{a}_2\right] \qquad [4.4]$$

where $\mathbf{a}_1$ and $\mathbf{a}_2$ are the real-space basis vectors corresponding to the diffraction vectors, $\mathbf{g}_1$ and $\mathbf{g}_2$. This is the same as for GPA of high-resolution images [HYT 98]. The strain tensor can then be obtained by numerical differentiation using the standard relations:

$$\varepsilon_{ij} = \frac{1}{2}\left(\frac{\partial u_i}{\partial x_j} + \frac{\partial u_j}{\partial x_i}\right) \qquad [4.5]$$

In a similar way, the local in-plane rigid body rotation, ω_{xy}, can be determined:

$$\omega_{xy} = \frac{1}{2}\left(\frac{\partial u_y}{\partial x_x} - \frac{\partial u_x}{\partial x_y}\right) \qquad [4.6]$$

where, for small rotations, the angle is in radians and anticlockwise positive. The more developed calculations for large deformations are not necessary here, though they are generally implemented in the practice.

A careful distinction needs to be made between deformation and strain. Strain in a mechanical sense is defined with respect to the stress-free, or undeformed, state. The reference lattice used to measure the deformation (i.e. $\mathbf{g}_1$ and $\mathbf{g}_2$) might not coincide with this undeformed lattice, in which case the deformation will not be the same as the strain. Second, in the presence of compositional variations that change the shape of the unit cell, the undeformed state will not be the same everywhere. Mixtures of strain and compositional variations are, therefore, the most difficult cases to analyze. Strictly speaking, DFEH, along with all microscopy techniques, only measures deformation.

4.3. Experimental requirements

DFEH experiments have the same general requirements as for conventional off-axis holography (see Chapter 1). The additional requirements concern the specimen and its orientation in the microscope. Before switching into holographic mode, the specimen should be oriented into a diffracting condition for the beam of interest. Ideally, this is a two-beam condition on a systematic row. A slight deviation from the exact Bragg angle is permitted in order to maximize the intensity in the diffracted beam. This will be necessary when the specimen thickness is not equal to a half extinction distance. The angle of rotation about the systematic row should be small, of the order of a degree: enough to reduce multiple scattering but not too large as to incline the viewing geometry adversely. At high angles from the zone axis, interfaces will become smeared in projection, a problem frequently encountered in CBED experiments. The overall aim is to obtain a dark-field image that has strong and uniform contrast over the field of view, and especially in the substrate that will serve as the reference. The difficulty is the question of specimen preparation.

Specimens of devices need to be prepared using FIB because of the site specificity. They need to be of uniform thickness and typically on, or around, a half multiple of the extinction distance of the diffracted beam of interest. Indeed, according to two-beam dynamic theory, and in an exact Bragg condition, the diffracted intensity is maximized at half the extinction distance, ξ_g, and in general $(n + \frac{1}{2})\xi_g$ where, n is an integer. The thicknesses to avoid are multiples of the extinction distance as the diffracted intensity is effectively zero.

For silicon at 200 kV, Table 4.1 gives the relevant thicknesses. If the {111} diffracted beams are to be used, the optimal sample thickness is about 120 nm, or $\frac{3}{2}\xi_g$. This value is to be preferred to 40 nm, $\frac{1}{2}\xi_g$, as this can be too thin for FIB specimen preparation. Higher multiples are possible but the intensity is reduced due to absorption. The other pair of beams often used for strain measurements are the

(220) beam combined with the (004), when perpendicular and parallel to the growth direction. Here, theoretically, the optimum thickness is at 250 nm for both beams but in practice, such specimens are too thick for accurate analysis. Finding a thinner optimal thickness is more difficult: an optimum thickness for the (220) beam is about 150 nm, whereas this is too close to the extinction distance for the (004) beam at 168 nm. We have found that a compromise around 120 nm is reasonable, shortly after the peak intensity for the (004) and before the peak in the (220). A deviation from the exact Bragg angle can be used to improve the diffraction intensity. A thickness of 120 nm can, therefore, be used for all the usual cases.

At the end of the day, there will never be a universal optimal thickness. Diffraction conditions will be modified by the local strains, and often, in a field of view, the materials present will not be the same: the obvious example being in the case of SiGe-recessed sources and drains. Neither do dynamical conditions ever fit exactly to two-beam theory. However, an obvious advantage of DFEH over HRTEM is that much thicker specimens can be used without adverse effect on the contrast. Specimen preparation is, therefore, easier and the effect of surface layers is smaller. Indeed, as shown in Table 4.1, DFEH can be carried out on HRTEM samples that are about 50 nm thick.

A more difficult problem for specimen preparation is bending such that the diffraction conditions are not uniform over the field of view, manifested by the presence of bend contours. Various means can be used to minimize bending but because the samples contain strain, some bending is inevitable.

Si at 200 kV	g_{111}	g_{220}	g_{004}
ξ_g (nm)	79	100	168
$\frac{1}{2}\xi_g$	40	50	85
$\frac{3}{2}\xi_g$	**119**	**149**	**252**
$\frac{5}{2}\xi_g$	158	**249**	338
$\frac{7}{2}\xi_g$	199	299	423

Table 4.1. *Extinction distances for silicon at an accelerating voltage of 200 kV and optimum thicknesses (in gray)*

Once good dark-field conditions have been obtained, the off-axis holography can be set up. Experiments can be carried out in high-resolution mode [COO 11] but for wide fields of view, the Lorentz mode is necessary. Unfortunately, most microscopes are not optimized for dark-field microscopy in the Lorentz mode. There are two principal problems: tilting the beam pre-specimen and using the objective aperture to select the diffracted beam. In the standard configuration, the relatively high angles of tilt in dark-field mode are compensated by the prefield of the objective lens. In the Lorentz mode, however, the objective is switched off, which limits the tilts that can be used. In the SACTEM-Toulouse, a Tecnai F20 equipped with an imaging aberration corrector (CEOS GmbH), we use the corrector as a Lorentz lens [HOU 08]. The maximum tilt is about 13 mrad, equivalent to the (220) beam in silicon at 200 kV. When acquiring dark-field holograms of the (004) planes, the diffracted beam will not, therefore, be exactly on the optic axis and some coma is inevitable. The second problem is the use of the objective aperture that is designed to be inserted in the diffraction plane of a microscope in standard mode. In the Lorentz mode, however, the diffraction plane does not usually coincide with this plane. The aperture is, therefore, blurred and generally the smallest objective apertures, which would be useful for increasing dark-field contrast, cannot be used.

These problems have been addressed in the design of the Hitachi I2TEM-Toulouse (in situ interferometry TEM), an HF3300 equipped with CFEG, double stage for Lorentz microscopy, imaging aberration corrector (CEOS Aplanator) and multiple biprisms. The second specimen stage is located just above the objective lens that now can serve as a powerful Lorentz lens coupled to the state-of-the-art aberration corrector. Holograms can be obtained with (008) diffracted beam with a resolution of 0.48 nm, compared with a typical resolution of 2–4 nm in a conventional Lorentz setup. The multiple biprisms allow more flexibility in choosing holographic fringe spaces and field of view, and Fresnel fringes can be eliminated [HAR 04]. Other practical aspects of the effect of a series of experimental parameters such as biprism voltage, exposure time, tilt angle and choice of the diffracted beam have been studied in detail on a silicon–germanium layer sample [BEC 11].

4.4. Strained silicon transistors with recessed sources and drains stressors

4.4.1. *Strained silicon p-MOSFET*

The example used to establish that the method works in practice was a dummy strained silicon channel p-MOSFET transistor [HYT 08]. Uniaxial strain is obtained in the silicon channel by the recessed sources and drains of $Si_{0.80}Ge_{0.20}$ grown in perfect epitaxy with the silicon (see Figure 4.1). A gate of polycrystalline silicon and gate oxide of silicon oxide complete the dummy device that lacks the subsequent steps of the process such as doping and contacting. The device is capped with a layer of silicon oxide and a thin layer of platinum (dark contrast) to protect the surface during specimen preparation. Samples were prepared using a modified H-bar process in the FIB to a thickness of about 120 nm.

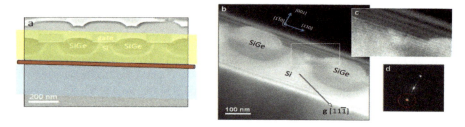

Figure 4.3. *Dark-field holography of a strained silicon p-MOSFET device: a) conventional bright-field image showing the gate, gate oxide and recessed sources and drains of SiGe (darker contrast), measurement (in yellow) and reference (in blue) regions, defined by the biprism position (in red); b) dark-field hologram using the [11-1] diffracted beam; c) enlargement showing narrow holographic fringes and larger Fresnel fringes; d) power spectrum showing side-band and mask radius*

The crystal was successively oriented into two-beam diffracting conditions to carry out the measurements. In each case, the biprism was positioned in such a way as to define a part of the substrate as the reference region (in blue in Figure 4.3) and the active region of the device where the strain is to be measured (in yellow). These regions have been presented on the bright-field image of the device so that the different elements can be seen. After applying a voltage (120 V in this case) to the biprism wire, the diffracted beam from these two regions is interfered to create the hologram. The hologram from the (11-1) diffracted beam is shown in Figure 4.3(b). The Si and SiGe regions are visible because they diffract, whereas the polycrystalline silicon, gate oxide and capping layers are invisible because they do not diffract. Indeed, close inspection of the hologram reveals that the holographic fringes are absent from these regions, with only the Fresnel fringes from the biprism visible (Figure 4.3(c)).

The phase of the hologram is then determined by the Fourier method. All dark-field hologram analysis was carried out using the patented software, HoloDark 1.0 [HOL]. The phase needs to be corrected for the distortions from the charge-coupled device (CCD) camera used to acquire the holograms. These can be eliminated by measuring the phase of a reference hologram in the vacuum, as for conventional electron holography (see Chapter 1), and subtracting it from the phase of the dark-field hologram. The drawback of this method is that the reference hologram is necessarily noisy and thus the overall effect is an increase in the noise. It is better to measure the distortions of the camera once and for all and then correct the phases automatically [HOL].

The resulting phase variations are very strong, up to several multiples of 2π. Again, we only see a meaningful phase in the diffracting regions of the specimen, the rest is noise. This is in stark contrast to electrostatic phase measurements (see Chapter 1), which are, orders of magnitude smaller. This means that the stringent specimen preparation requirements of uniform thickness are minimized somewhat compared with conventional holography. The effect of the Fresnel fringes is most visible in the substrate and is the principle remaining artifact.

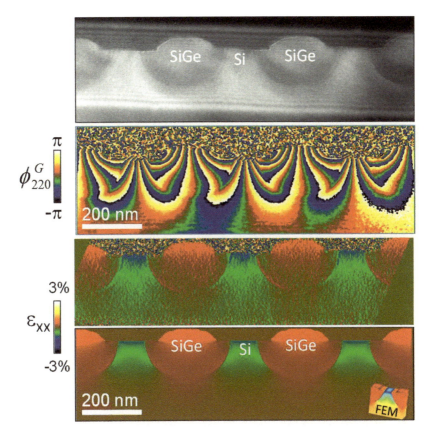

Figure 4.4. *strain analysis of dummy transistor: a) dark-field hologram for (220) diffracted beam; b) geometric phase determined from hologram; c) strain map determined from phase; d) simulated strain map by finite element method (FEM). Color scale identical for experimental and simulated strain*

The gradient of the phase (see equation [4.3]) then gives the local deformation of the lattice in the direction parallel to the diffraction vector [220], chosen as the *x*-axis. Gradients are determined numerically by subtraction of neighboring image points in the phase image. The result is shown in Figure 4.4(c). The silicon can be seen to be in compression in the channel region (green and blue on the color scale), increasing from the substrate to a maximum under the gate. The regions containing SiGe appear as areas of positive deformation (red colours), and reveal an important aspect of the measurements: deformation is always measured with respect to a reference, in this case the silicon substrate. Strain, in the mechanical sense, is defined with respect to the unstrained lattice. The deformation measured in the silicon is, therefore, equal to the strain, assuming the substrate is indeed fully relaxed. In regions of different composition, the misfit between silicon–germanium alloy and silicon needs to be subtracted in order to obtain the elastic strain, in our case 0.8%.

To determine the full strain tensor in two dimensions, two dark-field holograms need to be acquired. We chose two of the {111} diffracted beams (see Figure 4.5). The two resulting phase images then need to be aligned, which can be fastidious at times, as there is always a shift between the two holograms and is an essential part of the analysis software [HOL].

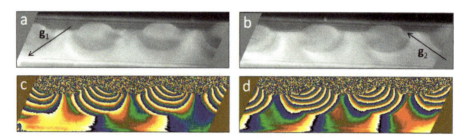

Figure 4.5. *determination of two-dimensional displacement field by dark-field holography: a) dark-field hologram for (−1−1−1) beam; b) dark-field hologram for (−1−11) beam; c) phase of (−1−1−1) beam; d) phase of (−1−11) beam*

The two phase images can then be used to determine the two-dimensional displacements (according to equation [4.4], and the components of the strain tensor determined (equations [4.5] and [4.6]). These are displayed in Figure 4.6.

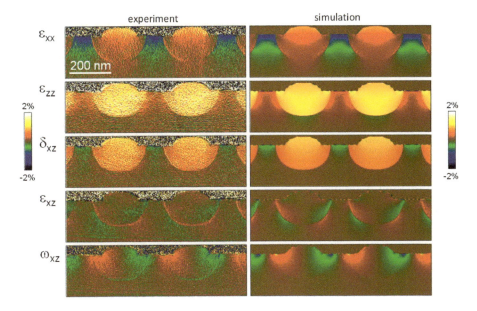

Figure 4.6. *Strain tensor for dummy transistor: a) experimental strain components and b) simulated strain components from finite element method*

The advantage of combining two dark-field holograms is that the full 2-D strain tensor information is obtained: not only the principal strain components along the transistor line (ε_{xx}) and the growth direction (ε_{zz}) but also the shear (ε_{xz}) and the local rigid body rotations (ω_{xz}). From only one set of lattice planes, it is impossible to distinguish between shear and rotation. We will see the use of this in a later example. The overall agreement between the experimentally determined strain maps and those coming from the modeling is excellent.

The artefacts from the Fresnel fringes are manifested in the fine set of fringes seen in the substrate. They are visible in all the strain components but the ε_{xx}. The reason for this is that the x-axis is almost perpendicular to the oscillating contrast that forms the Fresnel fringes, and hence a derivative in this direction is almost immune to their effect. One solution is to enlarge the hologram width and analyze only the central zone where their contrast is limited. Unfortunately, the hologram contrast diminishes with the overlap distance (see Chapter 1), which in turn reduces the signal-to-noise of the measured phase. A compromise is, therefore, necessary.

It can also be seen that at the interfaces between the SiGe and the silicon, particularly for the shear and the rotation, there appears to be a strong and localized strain contrast. This is also an artefact of the analysis and comes from two sources. The first is the difficulty of perfectly aligning the two holograms. Second, there is a difference in mean inner potential between the alloy SiGe and Si. The electrostatic phase will not, therefore, be uniform and there will be a (small) step in the phase at the interface. A large value for the derivative will, therefore, be registered right at the interface, which explains the localized nature of the contrast [HYT 11].

4.5. Thin film effect

At this point, we need to address an important issue inherent to any TEM investigation of strain: the thin-film effect. TEM foils are necessarily thin to allow electron beams to traverse them without too much loss in intensity. The thinning process introduces two free surfaces not present in the bulk sample that will minimize some of the stresses and strains present. In general, this minimization is typically 10% and is never more than 30% of the initial strain but needs, of course, to be estimated for quantitative measurements.

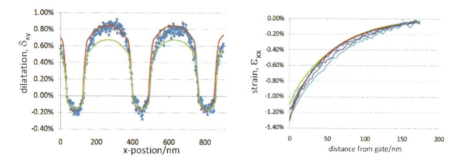

Figure 4.7. *Thin-film relaxation of measured strains: a) dilatation strain profiles across the transistor array; b) strain profiles from the substrate to the gate. Experimental curves (blue), plane-strain model (green) and bulk model (red)*

The upper and lower bound for measurements can be established by simulating the strain in an infinitely thin sample (plane stress conditions in elastic theory) and the infinitely thick, or bulk, sample (plane strain conditions). This was carried out for the p-MOSFET. Minimization depends very much on the strain component and the proximity with the surface, as shown in Figure 4.6(b). The dilatation reaches a maximum of 0.82% in the bulk simulation (in red) and 0.67% in the thin film (in green), which represents a reduction of 18%. However, the minimum values are identical. The experimental values (in blue) are closest to the bulk case because the specimen is relatively thick. For the lateral strain in the channel, ε_{xx}, the thin and the thick simulations give identical values in the substrate and gradually diverge as they approach the gate (at the surface) giving values of −1.1% (compression) and −1.3%, respectively. The experimental values lie between the two extremes and are closer to the bulk case at the gate.

4.6. Silicon implanted with hydrogen

Rather than dealing with the upper and lower bound for the strain, the analysis can be refined further. We will illustrate this with the study of hydrogen-implanted silicon. Ion implantation in Si is a standard process in modern technology. Depending on dose and implant species, it results in the distribution of point defects and/or complexes distributed all over the ion path or confined below the generated amorphous layer. During annealing, all these species precipitate and evolve in size in a way to reduce the elastic energy stored in the matrix. Thus, the mapping of strain as a function of depth is essential to understand and model the thermal evolution of such systems. The determination of strain using X-rays is not univocal and relatively insensitive to the shape of the strain distribution and even to its depth when the implanted layer is buried in the substrate. Figure 4.3 shows the strain mapping obtained in a Si wafer covered with 145-nm-thick SiO_2 and implanted with 32 keV, 1×10^{16} H^+/cm^2. These images show that the incorporation of H-related complexes in interstitial positions results in the build-up of an in-plane stress (with no relaxation) which relaxes in the zz-direction perpendicular to the wafer surface.

This sample was a uniform thin layer, so it was possible to use tripod polishing rather than FIB. In such cases, tripod polishing can give results as the thin foil is usually sufficiently uniform in thickness for DFEH. Indeed, pure mechanical polishing avoids the thick damage layers and Ga implantation due to the ion beam. Figure 4.8 shows the deformation map obtained from the (004) diffracted beam. The implantation has created a layer of silicon in tension in the [001] growth direction. No deformation was measured in the perpendicular direction. A profile was taken from the substrate to the surface, and averaged over a width of 100 nm (see Figure 4.8(c)).

The aim now is to a build a model that fits the experimental data taking into account the strain in the bulk sample and the thin film relaxation. Once a good fit is found, the free surfaces can be removed from the model and the bulk strain is determined. In this way, the relaxation can be corrected and a much better estimate for the initial strain can be obtained than comparing the upper and lower bounds as previously shown. In the case of hydrogen-implanted silicon, contrary to the p-MOSFET, no *a priori* model for the strain distribution exists. The experimental data were taken as a starting point for the bulk strain, nevertheless with a smooth profile corresponding to a log-normal distribution. The amplitude of this distribution was then varied until a good fit could be found (green curve on Figure 4.8). This model naturally included the relaxation with two free surfaces reflecting the true specimen thickness of about 120 nm (measured from the extinction contours). The free surfaces were then removed and the strain distribution in the bulk sample (red curve) was determined. This is the methodology that has been established for the following examples.

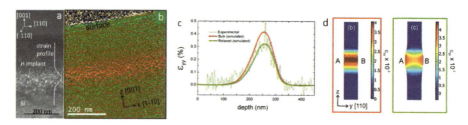

Figure 4.8. *Hydrogen-implanted silicon and the thin film effect: a) conventional dark-field image; b) strain map determined by DFEH from (004) diffracted beam; c) strain profiles: experimental (gray), simulated thin film (green) and bulk (red); d) cross-section of model: bulk (red box), thin film (green box)*

4.7. Strained silicon n-MOSFET

To increase the mobility of electron carriers in an n-type MOSFET, the silicon channel should be in tension. To apply uniaxial stress as for the p-MOSFET, the lattice parameter in the sources and drains should be smaller than silicon. This can be achieved by doping silicon with a small amount of carbon. There is a limit, however, as to the quantity of carbon that can be incorporated substitutionally. With increasing doses, carbon will be incorporated as interstitials, which hardly modifies the lattice parameter, and ultimately will precipitate in the form of SiC, again with limited effect on the lattice parameter. This sequence of events has been quantified by measuring the strain in silicon-doped thin films [CHE 09]. Nevertheless, there is increased uncertainty concerning the injected strain when using the recessed sources and drains with this system and experimental verification is necessary.

Figure 4.9 shows the analysis of a dummy n-type MOSFET with recessed silicon, nominally doped with 1% carbon [HUE 09]. Dark-field holography using the (220) diffracted beam shows that indeed the silicon channel region is in tension. The lattice parameter in the region of Si:C can also be seen to be slightly smaller than the silicon. To quantify this, profiles were taken vertically from the substrate to the gate in the channel region and between the gates (Figure 4.9(d)). The profiles were averaged over the width of the channel (65 nm) to improve the signal-to-noise ratio. The agreement between the five channels in the field of view is remarkable. The standard deviation between the curves is only 0.02%, which is our estimation of the precision of the technique. The diminution of the lattice parameter in the Si:C is small and reaches a maximum of about 0.15% at the surface. The value for the strain measured just under the gate is 0.58 ± 0.02%.

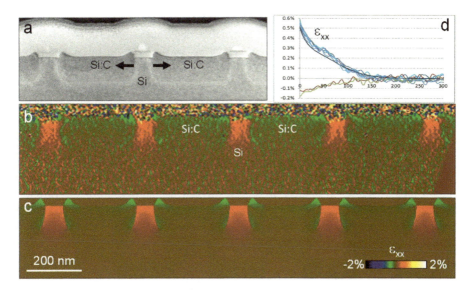

Figure 4.9. *Analysis of an n-MOSFET with recessed source and drain of carbon-doped silicon: a) bright-field image; b) DFEH map of the horizontal strain component, ε_{xx}, showing the channels to be in tension (red); c) finite element method modeling with identical color scale; d) strain profiles from substrate to gate in the channel (blue), between channels (red, green) and best-fitting model (black)*

These strains were compared with finite element method modeling using the same methodology as the p-MOSFET. Assuming the full 1% of carbon was incorporated substitutionally, it gave values of strain that were systematically too high. We, therefore, varied the carbon content until a good fit was found (see Figure 4.9(d)). The best fit was obtained for 0.75% carbon on substitutional sites. This value is in excellent agreement with the results from the analysis of thin films. There it was found that for 1% of total carbon composition, between 20% and 25% of the carbon was on interstitial sites and only 70–75% on substitutional sites.

This example shows the value of carrying out direct experimental verification of the strain in the channel and also the very high precision of the technique. The level of strain in the n-MOSFET is much lower than the p-MOSFET due to the diminishing returns of injecting carbon.

4.8. Understanding strain engineering

DFEH, as we have shown, is a powerful technique for characterizing strain in devices. The measurements could be reduced to determine the value of strain just under the gate in the channel region. However, this would be missing the real potential of the technique. The wealth of information provided by the mapping of strain components allows another more interesting possibility that is: to understand the different processes making up strain in devices.

Strain can be introduced into a device in a number of ways, both intentionally and unintentionally (see Figure 4.10). While stress liners and alloyed sources and drains are intended to strain the channel region, contacting may also cause stress. The local geometry will also determine how strain can be relaxed or concentrated in certain areas. Neither is it clear how stresses will accumulate from the different causes, whether they will be simply additive or some more complex processes will be involved. It is, therefore, interesting to work on devices at different stages of the production process and using different combinations of strain engineering.

Recessed source and drains are not the only way to engineer strain in the channel. Capping layers (typically, contact etch stop layers (CESL)) can also provide the necessary stresses. Although it is established that the covering of an n-MOS transistor with a CESL layer of Si_3N_4 results in an increased mobility of the electrons in the channel region, the reasons for this effect have not been directly elucidated. A simplified example of such a device is shown in Figure 4.10(b). The dummy transistor consists of a polysilicon gate, gate oxide (silicon oxide) and stress liner (silicon nitride). DFEH reveals the strain caused by this structure on the silicon substrate. To make sense of the strain field, modeling would have to be performed which includes all of these three features, each of which can produce strain. It is therefore interesting to work on dedicated structures on which this effect can be more easily observed and understood.

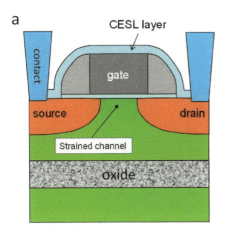 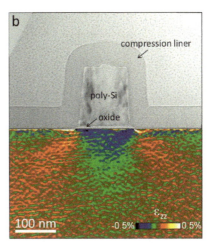

Figure 4.10. *Analyzing strain engineering: a) schematic view of a strained silicon channel transistor showing different possible sources of strain and geometry; b) dummy transistor (bright-field image in gray-scale) using a compression liner to strain the substrate (strain map determined by DFEH in color)*

4.9. Strained silicon devices relying on stressor layers

Figure 4.11 shows a test structure for the study of the strain induced by CESL. It consists of a comb-like periodical array of etched silicon trenches a few hundred nanometers deep and about 50–100 nm wide. The Si is covered by a capping layer of a few tens of nanometers thick of Si_3N_4 followed by silicon oxide. Differences in thermal expansion coefficients and/or exodiffusion of hydrogen from the deposited layer result in high levels of stress in the nitride layer after processing, which will in turn deform the silicon.

The thinned specimen shows a feature common to many samples prepared by FIB: shadowing. The ion beam was used on the specimen only from the "top-side", which means it was directed toward the surface of the specimen. The oxide between the silicon "prongs" of the comb provides less resistance to the beam and grooves are formed in the substrate immediately below (see Figure 4.11(a)). The thickness of the specimen is therefore uneven, the effects of which will be seen in the subsequent analysis. Back-side thinning provides a solution to this problem, but is more difficult to put into practice.

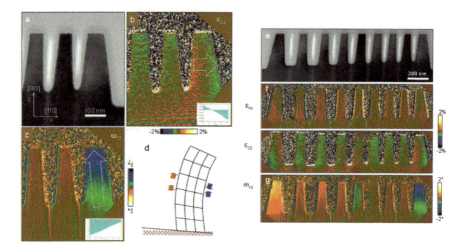

Figure 4.11. CESL test structure analyzed by DFEH: a) conventional bright-field TEM image of etched silicon structure coated with thin Si3N4 layer, brighter lines in substrate are the results of curtaining in the FIB specimen preparation; b) strain map, ε_{zz}, of right-end elements (see framed box in (b)), green colors indicate compressive strains and red in tension; c) lattice rotation map, ω_{xz}, showing progressive bending of right-hand element; d) montage of results obtained from three holograms to show strain components, ε_{xx}, ε_{zz} and ω_{xz}, mapped across 1.5 μm of device. Strains are tensile in ε_{xx}, varying from 0.0% to 0.1%, and compressive in ε_{zz}, varying from −0.09% to −0.29%. Note the bending of opposite bending of the end elements in ω_{xz}; e) horizontal profile of ε_{zz} plotted as indicated by arrow in (b); f) schematic diagram showing bending of end element; g) vertical profile of ω_{xz} plotted as indicated by arrow in (c)

Dark-field holograms were acquired using two of the {111} diffracted beams and the strain tensor determined (see Figures 4.11(b) and (c)). In general, the Si lattice in the vertical elements is negatively strained (compression) with values lying in the 0.1–0.3% range (green colors). However, the situation in the right end element is significantly different. In this region, the strain changes from positive to negative from the left to right (see profile, Figure 4.11(b)). It is difficult to understand this strain distribution by only considering the deformation in one direction, for example if we had studied just one diffracted beam. Fortunately, we also have access to the in-plane rotation of the crystal with respect to the perfect Si lattice (Figure 4.11(c)). From the center of the structure toward its edge, the elements tend to rotate, the rotation being particularly pronounced at the end of the structure. Interestingly, this rotation angle increases almost linearly from the bottom to the top. This information allows us to deduce the underlying characteristics of the strain state of the Si lattice in such a structure.

A schematic diagram representing the lattice in the element at the end of the comb-like structure is shown in Figure 4.11(d). The Si crystal is elastically bent toward the right side. For this reason, the rotation angle increases from the bulk and toward the surface, just like a macroscopically bent beam. As a result, the lattice is in vertical compression on the right side and in vertical tension on the left side. The cause of the lattice rotations can be understood intuitively as a way of partially relieving the compressive stresses applied by the CESL on the overall structure by bending at the extremities. Moreover, a detailed examination of the bright-field images reveals that the liner thickness is significantly different on the left and right side of the end-most element. Finite element modeling confirms that the disequilibrium in the compressive forces provides the torque necessary for the bending. This example shows the need to determine an ensemble of strain components in order to understand the way strain acts even in such relatively simple structures.

4.10. 28-nm technology node MOSFETs

Figure 4.12 shows a complete n-MOSFET device from the 28-nm technology node (courtesy of STMicroelectronics) including contacting, stress liners and doping. The grooving in the silicon substrate is due to the FIB preparation that was carried out on front side. Dark-field holography was used to measure the strain, notably the component parallel to the source-drain direction (Figure 4.12(b)). The channel region can be seen to be in tension, as required to increase the mobility of electrons in n-MOS devices. The raising of the channel region allows more leverage to stretch the silicon with the stress liner. Another feature is also visible in the strain map: the large scale compression of the whole region containing the transistors (see the dotted line). The global geometry of the device structure can, therefore, interfere with the localized injection of strain. In this case, the large-scale compression competes with the local tension in the channel. This example also illustrates the necessity to measure strain over a wide field of view while maintaining nanometer resolution. If, for example, strain had only been mapped locally, with HRTEM for example, the reference region for the measurements would in fact be in compression producing a systematic error.

A p-MOSFET from a similar series has also been analyzed by DFEH (Figure 4.13). In this case, the channel region is made up of a thin layer of SiGe alloy, revealed by the positive deformation (in red) on the strain map of the vertical component, ε_{zz}. The x-component, ε_{xx}, also shows a positive deformation that is at first unexpected. In addition, the deformation varies smoothly from the gate to the substrate and is not sharply delineated like the other strain components.

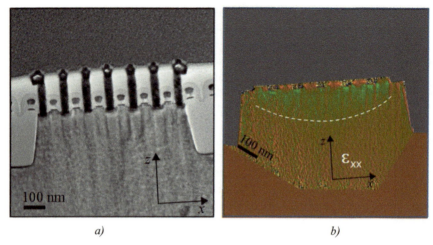

Figure 4.12. *n-MOSFET from the 28-nm technology node: a) conventional bright-field image showing the architecture; b) strain map, ε_{xx}, parallel to the source-drain direction determined by DFEH. Surface grooving from the FIB preparation produce artefacts in the strain as vertical strips. Dotted line indicates the region of large-scale compression*

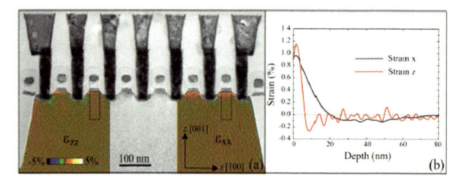

Figure 4.13. *p-MOSFET from the 28-nm technology node and SiGe channel: a) bright-field image (gray-scale) and DFEH strain maps of the vertical component, ε_{zz}, (color scale on left) and horizontal component ε_{xx} (color-scale on right); b) strain profiles from gate to substrate (region boxed on (a))*

Figure 4.14(a) shows the expected configuration for a strained SiGe layer on Si on a full sheet. The ε_{zz} component is positive in the layer whereas the ε_{xx} component is zero everywhere. The reason is that the SiGe layer is forced to take the lattice parameter of the silicon substrate. The layer is, therefore, in bi-axial *compression*. As always, the strain map is a measure of the deformation with respect to a reference lattice, and not necessarily the relaxed state. In this case, the relaxed lattice parameter of the SiGe layer is larger than the silicon, so it is compressed to fit the Si lattice parameter (which serves as the reference). In the z-direction, the lattice can relax; indeed, due to the Poisson effect in reaction to in-plane compression, it is even larger than the relaxed state. However, in the transistor, the local geometry is different (see Figure 4.14(b)). The fact that the channel region is raised allows lateral relaxation of the strain. The SiGe layer can expand to relieve the compressive strain. We can see this particularly at the edges of the channel. Relaxation of the SiGe layer will be transmitted to the substrate, creating tension in the silicon that will gradually disappear further into the substrate (Figure 4.14(c)).

Overall, the channel remains in compression, which was necessary to boost the hole mobility, but is less than predicted from full sheet simulations. Only by carrying out the measurements, it is possible to quantify this effect.

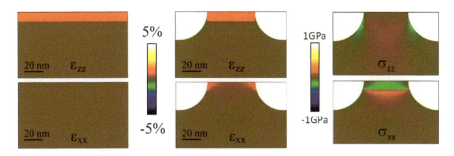

Figure 4.14. *FEM modeling of strained layer of SiGe on Si: a) full sheet configuration; b) raised channel geometry; c) stresses in raised channel geometry*

4.11. FinFET device

New 3D geometries and silicon-on-insulator (SOI) devices are a challenge for characterization techniques. An important example is the so-called FinFET. The architecture studied here (see Figure 4.15(a)) combines both the 3D aspect and the presence of an oxide layer separating the active part of the transistor from the substrate. A narrow "fin" of silicon is suspended above an oxide layer. The gate is wrapped around the fin on three sides for precise control of the current passing along the fin from source to drain. In addition, the gate is covered with TiN to introduce stress to the channel region to increase mobility (Figure 4.15(b)). Currently, it is not possible to measure strain tomographically in three dimensions, so a dummy FinFET device was created with extended fin length. It was then possible to cut a slice through the structure by FIB.

The resulting specimen is shown in Figure 4.15(c). There are three challenges: the region of interest is relatively small, the reference region is relatively far, and there may be disorientation between the substrate and region of interest. For DFEH to work, the substrate and the region of interest must diffract simultaneously. However, it is not necessary that the two be perfectly aligned into a Bragg condition. A compromise can be found and the crystal oriented until both regions have contrast for the diffracted beam chosen for the analysis. Again, a wide field of view and nanometric spatial resolution are necessary to capture the details in the fin and supply a suitable reference. We achieved this for two {111} diffracted beams. In this case, it was easy to align the two holograms and calculate the strain field (see Figure 4.15(d)).

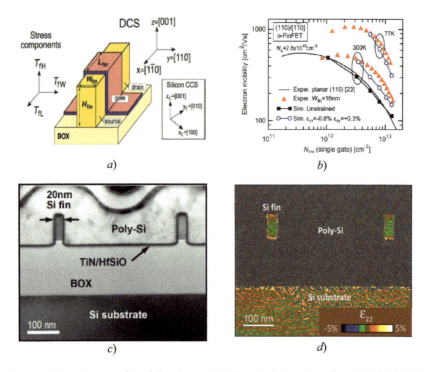

Figure 4.15. *Strain and mobility in n-FinFETs: a) Schematic of a (110)/[1-10] FinFET obtained with a standard Manhattan layout on a (100) wafer, device coordinate system (DCS) and crystal coordinate system (CCS) also shown; b) measured and simulated electron mobility in n-FinFETs at 300 and 77 K (circled); c) bright-field image of FIB-prepared lamella; d) DFEH measurement of strain (non-diffracting region grayed out)*

The strain is relatively uniform within each fin and values measured from different structures range from –0.4% to –0.8% strain (compression). These values were used in the modeling of the electrical characteristics of the corresponding device and compared with the experimental values measured on the full device [CON 11].

4.12. Conclusions

DFEH is a fully quantitative technique for the measurement of strain in nanostructures and devices. The analysis is straightforward and gives the strain components directly that can be compared with modeling. DFEH is, in particular, adapted to epitaxial systems as the substrate can act as a reference area. The technique has already been applied successfully to a number of systems, from the MOSFET and FinFET devices shown here [HYT 08, HUE 09, CON 11] and similar strained silicon devices [COO 10], to strained layers [BEC 11], misfit dislocations therein [HAR 10] and quantum dots [COO 11]. The technique can be powerfully combined with conventional holography to provide a complete study of strain and dopants in devices [COO 11].

4.13. Bibliography

[ACO 06] ACOSTA A., SOOD S., "Engineering strained silicon: looking back and into the future", *IEEE Potentials*, vol. 25, pp. 31–34, 2006.

[ANT 06] ANTONIADIS D.A., ABERG I., NI CHLEIRIGH C., NAYFEH O.M., KHAKIFIROOZ A., HOYT J.L., "Continuous MOSFET performance increase with device scaling: the role of strain and channel material innovations", *IBM Journal of Research and Development*, vol. 50, pp. 363–376, 2006.

[BEC 11] BECHE A., ROUVIERE J.L., ARNES J.P., COOPER D., "Dark field electron holography for strain measurement", *Ultramicroscopy*, vol. 111, pp. 227–238, 2011.

[CHE 09] CHERKASHIN N., HÝTCH M.J., HOUDELLIER F., HÜE F., PAILLARD V., CLAVERIE A., GOUYÉ A., KERMARREC O., ROUCHON D., BURDIN M., HOLLIGER P., "On the influence of elastic strain on the accommodation of carbon atoms into substitutional sites in strained Si:C layers grown on Si substrates", *Applied Physics Letter*, vol. 94, p. 141910, 2009.

[CLE 04] CLEMENT L., PANTEL R., KWAKMAN L.F.T., ROUVIERE J.-L.,"Strain measurements by convergent-beam electron diffraction: the importance of stress relaxation in lamella preparations", *Applied Physics Letter*, vol. 85, pp. 651–653, 2004.

[CON 11] CONZATTI F. *et al.*, "Investigation of strain engineering in FinFETs comprising experimental analysis and numerical simulations", *IEEE Transactions on Electronic Devices*, vol. 58, pp. 1583–1593, 2011.

[COO 10] COOPER D., BÉCHÉ A., HARTMANN J.-M., CARRON V., Rouvière J.-L., "Strain evolution during the silicidation of nanometer-scale SiGe semiconductor devices studied by dark field electron holography", *Applied Physics Letter*, vol. 96, p. 113508, 2010.

[COO 11] COOPER D., ROUVIÈRE J.-L., BÉCHÉ A., KADKHODAZADEH S., SEMENOVA E.S., YVIND K., DUNIN-BORKOWSKI R.E., "Quantitative strain mapping of InAs/InP quantum dots with 1nm spatial resolution using dark field electron holography", *Applied Physics Letter*, vol. 99, p. 261911, 2011.

[COO 11] COOPER D., DE LA PENA F., BECHE A., ROUVIERE J.-L., SERVANTON G., PANTEL R., MORIN P., "Field mapping with nanometer-scale resolution for the next generation of electronic devices", *Nano Letters*, vol. 11, p. 4585, 2011.

[FOR 06] FORAN B., CLARK M.H., LIAN G., "Strain measurement by transmission electron microscopy", *Future Fab International*, vol. 20, pp. 127–129, 2006.

[GHA 03] GHANI T., ARMSTRONG M., AUTH C. et al., "A 90 nm high volume manufacturing logic technology featuring novel 45 nm gate length strained silicon CMOS transistors", *IEDM Techinal Digest (IEEE International)*, pp. 978–980, 2003.

[GPA 11] Phase 3.0 (HREM Research Inc.), A plug-in for the image processing package DigitalMicrograph (Gatan), available at http://www.hremresearch.com, 2011.

[HAR 04] HARADA K., TONAMURA A., TOGAWA Y., AKASHI T., MATSUDA T., *Applied Physics Letter*, vol. 84, p. 3229, 2004.

[HAR 10] HARTMANN J.-M. et al., "Fabrication, structural and electrical properties of compressively strained Ge-on-insulator substrates", in L. SANCHEZ, W. VAN DEN DAELE, A. ABBADIE, L. BAUD, R. TRUCHE, E. AUGENDRE, L. CLAVELIER, N. CHERKASHIN, M. HYTCH, *Semiconductor Science and Technology*, vol. 45, p. 075010, 2010.

[HOL] HOLODARK 1.0 software (HREM Research Inc.), A plug-in for the image processing package DigitalMicrograph 3.5 (Gatan), available at http://www.hremresearch.com.

[HOU 06] HOUDELLIER F., ROUCAU C., CLEMENT L., ROUVIERE J.-L., CASANOVE M.-J., "Quantitative analysis of HOLZ line splitting in CBED patterns of epitaxially strained layers", *Ultramicroscopy*, vol. 106, pp. 951–959, 2006.

[HOU 08] HOUDELLIER F., HŸTCH M.J., HÜE F., SNOECK E., in HAWKES P.W. (ed.), "Aberration correction with the SACTEM-Toulouse: from imaging to diffraction", *Advances in Imaging and Electron Physics*, Chapter 6, Elsevier, Amsterdam, vol. 153, pp. 1–36, 2008.

[HÜE 05] HÜE F., JOHNSON C.L., LARTIGUE-KORINEK S., WANG G., BUSECK P.R., HŸTCH M.J., "Calibration of projector lens distortions", *Journal of Electron Microscopy*, vol. 54, pp. 181–190, 2005.

[HÜE 08] HÜE F., HŸTCH M.J., BENDER H., HOUDELLIER F., CLAVERIE A., "Direct mapping of strain in a strained-silicon transistor by high-resolution electron microscopy", *Physics Review Letter*, vol. 100, p. 156602, 2008.

[HÜE 09] HÜE F., HŸTCH M.J., HOUDELLIER F., BENDER H., CLAVERIE A., "Strain mapping of tensiley strained silicon transistors with embedded $Si_{1-y}C_y$ source and drain by dark-field holography", *Applied Physics Letter*, vol. 95, p. 073103, 2009.

[HŸT] HŸTCH M.J., HOUDELLIER F., HÜE F., SNOECK E., International patent application No PCT/FR2008/001302, 2008.

[HŸT 98] HŸTCH M.J., SNOECK E. KILAAS R., "Quantitative measurement of displacement and strain fields from HREM micrographs", *Ultramicroscopy*, vol. 74, pp. 131–146, 1998.

[HŸT 03] HŸTCH M.J., PUTAUX J-L., PÉNISSON J-M., "Measurement of the displacement field around dislocations to 0.03Å by electron microscopy", *Nature*, vol. 423, pp. 270–273, 2003.

[HŸT 08] HŸTCH M.J., HOUDELLIER F., HÜE F., SNOECK E., "Nanoscale holographic interferometry for strain measurements in electronic devices", *Nature*, vol. 453, pp. 1086–1089, 2008.

[HŸT 11] HŸTCH M.J., HOUDELLIER F., HÜE F., SNOECK E., "Dark-field electron holography for the measurement of geometric phase", *Ultramicroscopy*, vol. 111, pp. 1328–1337, 2011.

[ITR 11] ITRS (ITRS, Austin, USA) www.itrs.net, International Technology Roadmap for Semiconductors 2011 Edition.

[JAC 06] JACOBSEN R.S., ANDERSEN K.N., BOREL P.I., FAGE-PEDERSEN J., FRANDSEN L.H., HANSEN O., KRISTENSEN M., LAVRINENKO A.V., MOULIN G., OU H., PEUCHERET C., ZSIGRI B., BJARKLEV A. "Strained silicon as a new electro-optic material", *Nature*, vol. 441, pp. 199–202, 2006.

[JOH 08] JOHNSON C.L., SNOECK E., EZCURDIA M., RODRÍGUEZ-GONZÁLEZ B., PASTORIZA-SANTOS I., LIZ-MARZÁN L.M., HŸTCH M.J., "Effects of elastic anisotropy on strain distributions in decahedral gold nanoparticles", *Nature Materials*, vol. 7, pp. 120–124, 2008.

[LEE 05] LEE M.L., FITZGERALD E.A., BULSARA M.T., CURRIE M.T., LOCHTEFELD A., "Strained Si, SiGe, and Ge channels for high-mobility metal-oxide-semiconductor field-effect transistors", *Journal of Applied Physics*, vol. 97, p. 011101, 2005.

[PAR 06] PARTON E., VERHEYEN P., "Strained silicon – the key to sub-45 nm CMOS", *III-Vs Review*, vol. 19, pp. 28–31, 2006.

[THO 06] THOMPSON S.E., SUN G.Y., CHOI Y.S., NISHIDA T. "Uniaxial-process-induced strained-Si: extending the CMOS roadmap", *IEEE Transactions on Electron Devices*, vol. 53, pp. 1010–1020, 2006.

[USU 05] USUDA K., NUMATA T., IRISAWA T., HIRASHITA N., TAKAGI S., "Strain characterization in SOI and strained-Si on SGOI MOSFET channel using nano-beam electron diffraction (NBD)", *Materials Science and Engineering: B*, vol. 124, pp. 143–147, 2005.

[ZHA 06] ZHANG P., ISTRATOV A.A., WEBER E.R., KISIELOWSKI C., HE H.F., NELSON C., SPENCE J.C.H., "Direct strain measurement in a 65 nm node strained silicon transistor by convergent-beam electron diffraction", *Applied Physics Letter*, vol. 89, p. 161907, 2006.

Chapter 5

Magnetic Mapping Using Electron Holography

5.1. Introduction

The recent know-how in manipulating the spin of electrons combined with their charge to transfer information has opened new routes in so-called "spintronic" devices that already have found applications in data storage and magnetic sensors hyper-frequency generators and have high potentialities in logics devices [CHA 07]. Similar to microelectronic compounds in which the size reduction induces remarkable properties, magnetic nanosystems exhibit various potential opportunities. Magnetic materials (3D types and magnetic semiconductors) are, therefore, largely used in the microelectronic industries in hybrid systems combined with "normal" silicon devices. Likewise, microelectronic devices that require the knowledge of their structural and chemical properties and also of the local electrical potential, strain field, etc., the local magnetic properties of spintronic systems need to be determined to optimize their use as well. Electron microscopy has proved to be an efficient tool for such goals and transmission electron microscopy (TEM) has been used by several groups worldwide in the analysis of magnetic nanomaterials and systems [TON 92, RAF 98, RAF 00, RAF 02, McC 97, McC 98, BEL 03, SNO 03, SNO 08].

As presented in Chapter 1, the phase of an electron wave interacting with a sample is sensitive to the electrostatic, magnetic and strain fields within the specimen and possibly around it. However, in a conventional TEM experiment, only

---

Chapter written by Etienne SNOECK and Christophe GATEL.

the spatial distribution of beam intensity is recorded and all information about the phase shift of the high-energy electron wave are lost. Electron holography is a powerful interferometry technique allowing the phase shift of the electron wave to be recovered. Because the phase shift is sensitive to the in-plane components of the magnetic induction, electron holography can, therefore, provide quantitative information on the magnetic properties of ferromagnetic materials and devices such as domain wall configuration, in-plane magnetization and magnetization reversal processes with a resolution of few nanometers in conventional cold- or Schottky-FEG microscopes and of few ångströms in dedicated Lorentz corrected TEMs.

In this chapter, we will detail the capabilities of electron holography for the study of the magnetic properties of magnetic (nano)materials. We will give the specific experimental constraints and issues to overcome the study of magnetic materials and describe the hologram analysis procedures to get quantitative information. We explain the method for studying the magnetic properties of two neighboring Fe nanocubes. We will then illustrate the performance of the electron holography method for the study of the magnetic properties of a FePd (L10) epitaxial thin film that can be used for perpendicular magnetization recording (PMR).

In the last section, we will draw out the prospect of this technique for use in future years.

5.2. Experimental

The phase shift of an electron beam that has interacted with an electromagnetic field is sensitive both to the electrostatic potential "V" and to the components of the magnetic induction perpendicular to the beam direction "$B_\perp$" and integrated along the beam path.

The measured phase shift is derived from the Aharonov–Bohm expression giving the phase shift of charged particles interacting with an electrostatic potential and a magnetic vector potential [AHA 59]:

$$\phi(x) = C_E \int V(x,z)\,dz - \frac{e}{\hbar} \int A_z(x,y)\,dz \qquad [5.1]$$

$$\phi(x) = C_E \int V(x,z)\,dz - \frac{e}{\hbar} \iint B_\perp(x,z)\,dx\,dz \qquad [5.2]$$

where:

$$C_E = \frac{2\pi}{\lambda} \frac{E_k + E_0}{E_k(E_k + 2E_0)} \quad [5.3]$$

– z is the direction of the incident beam.

– x is a direction of the plane perpendicular to beam direction.

– A_z is the component of the magnetic vector potential "A" along the beam direction.

– $B_\perp$ is the component of the magnetic induction perpendicular to both x and z.

– V is the electrostatic potential.

– λ is the electron wavelength.

– E_0 is the rest mass energy of the electrons.

– E_k is the kinetic energy of the electrons.

The total phase shift can be separated into two terms: the electrostatic contribution to the phase shift "ϕ^E", and the magnetic contribution "ϕ^M" (in the notation of M. Hÿtch et al. [HYT 11]).

$$\phi^E(x) = C_E \int V(x,z)\,dz \quad [5.4]$$

$$\phi^M(x) = -\frac{e}{\hbar} \iint B_\perp(x,z)\,dx\,dz \quad [5.5]$$

– The electrostatic phase shift "ϕ^E" is the sum of the phase shift due to the specimen called mean inner potential (MIP), $V_0(x,z)$ and of possible additional local electrical potential, $V_E(x,z)$, related to the presence of dopants or charges, etc., such as the electrostatic potential obtained in PN junctions, charged particles and doped regions in transistors (see Chapter 1). The MIP contribution is always present whatever the specimen and only depends on the nature of the atoms constituting the material whereas the second electrostatic term is generally not expected in a non-doped system. It, however, may appear if the specimen charges under the electron beam irradiation.

– The magnetic phase shift "ϕ^M" can be understood as the result of the Lorentz force that deflects the electron beam in a direction perpendicular both to the incident beam direction and the magnetic field direction. This implies that only the components of the magnetic field that are perpendicular to the electron beam can be recorded in electron holography.

These points induce two constraints that need to overcome in order to perform quantitative measurements of magnetic properties in electron holography: magnetic contribution needs to be separated from the electrostatic part and the magnetic information obtained from electron holography experiments is the projection along the electron path of the in-plane induction.

5.2.1. *The Lorentz mode*

In an electron microscope, lenses are magnetic lenses and the sample is located within the pole piece of the objective lens that creates, when excited, an important magnetic induction parallel to the optic axis of the microscope (i.e. along z direction). The induced magnetic field can reach 2 T in a 200 kV TEM (see Figure 5.1) and even 3 T in a 300 kV microscope.

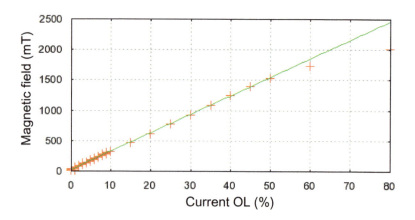

Figure 5.1. *Variation in the magnetic induction within the pole piece of the objective lens as a function of the electric current passing through the coils of the objective (data recorded on the SACTEM Tecnai F20 (FEI) 200 KV microscope in CEMES)*

Most of the magnetic materials or devices embedded in such a high magnetic field will have their magnetization switching along the direction of the applied field, that is along the z direction, and no magnetic signal can be recorded in electron holography (*neither in Lorentz microscopy*).

This implies that to measure any magnetic properties of a given magnetic material in TEM, the objective lens has to be switched off and the optic configuration of the microscope has to be set up in such a way that other lens located below the sample can be used to create the image. Such TEM configuration is called the "Lorentz mode" where the sample is located in a field-free environment where no magnetic field of the objective lens is expected to modify its magnetic configuration, the specimen being in a remnant magnetic state.

The electron holography setup in the Lorentz mode is, however, similar to the one described in Figure 1.2 in Chapter 1. The illumination has to be set highly astigmatic in a direction perpendicular to the biprism, the voltage applied to the biprism and the exposure time have to be adjusted in such a way as to optimize the signal-to-noise ratio (SNR) and the contrast of the hologram while recording a sufficient number of fringes on each pixel of the CCD detector for an optimal spatial resolution.

As described in Chapter 1, a reference hologram has to be recorded in the vacuum far from any long-range electrostatic and/or magnetic stray fields to remove, in a post-experiment computing procedure, artifacts coming from the Fresnel fringes due to the biprism and hologram fringes distortions coming from the camera and lens aberrations. The phase and amplitude images are recovered performing a Fourier analysis of the hologram as explained in Chapter 1.

The main difference between electron holography experiments performed in "normal" mode (i.e. objective lens on) and in Lorentz mode is that the field of view and the overlap area are much smaller in "normal" mode (typically few tens of nanometers) whereas in Lorentz mode, areas could reach few microns. For further details about optimum TEM configuration for electron holography experiment, refer to Jan Sickmann *et al.*'s paper [SIC11] that presents a detailed study of optical TEM configurations to have access to a continuous range of the field of view size and spatial resolution from few microns to few nanometers.

5.2.1.1. Is Lorentz mode really a field-free environment?

One has to be aware that even when switching the objective lens off and setting the microscope in its Lorentz mode, the sample may not be in a real field-free environment as the objective lens could present a remnant field. The later may be high enough (few hundreds of Oersted) to modify the sample magnetic configuration in the case of a soft magnetic material such as permalloy, for instance. In addition, even if the magnetic field of the objective lens is indeed zero in the middle of the pole pieces, it has been shown that it may be non-zero in the direction perpendicular to the optic axis. In particular, the sample can experience a magnetic field up to few tens of Oersted when inserting the sample into the column (R. Dunin Borkowski, private communication).

5.2.2 The "ϕ^E" problem.

As the phase shift $\phi(x)$ measured within magnetic sample always contains the electrostatic $\phi^E(x)$ and the magnetic $\phi^M(x)$ contributions, strategies have been carried out to separate the two contributions and get the magnetic phase or the electrostatic part only. There are basically two experimental ways for succeeding in separating these two contributions. They both consist of taking two holograms (1) and (2) between, which, the sign of $\phi^M(x)$ is reversed.

$$\phi_1(x) = \phi^E(x) + \phi^M(x) \text{ and } \phi_2(x) = \phi^E(x) - \phi^M(x)$$

The sum and the difference of the two phase images then gives twice the electrostatic contribution and twice the magnetic contribution, respectively.

$$\phi^M(x) = \frac{\phi_1(x) + \phi_2(x)}{2} \qquad [5.6a]$$

and

$$\phi^E(x) = \frac{\phi_1(x) - \phi_2(x)}{2} \qquad [5.6b]$$

1) The first way of reversing the sign of $\phi^M(x)$ is to reverse the magnetization of the sample between experiments (1) and (2) in two opposite directions. In such case:

$$\phi_1^M(x) = \frac{e}{\hbar} \iint B_\perp(x,z) dxdz$$

$$\phi_2^M(x) = \frac{e}{\hbar} \iint -B_\perp(x,z) dxdz = -\phi_1^M(x)$$

2) The second way is related to the fact that $\phi^M(x)$ is a geometric phase that is reversed in sign when reversing the time. In our experiment, reversing the time corresponds to reversing the direction of the incident electron beam and then switching the specimen upside down between the two experiments.

Each separating method presents some advantages and drawbacks. But both of them need to record at least two holograms (four with the reference holograms).

5.2.2.1. *Saturating the sample magnetization in two opposite directions*

The high magnetic field created by the objective lens coils when they are switched on can be used to align the magnetization in a given direction. To do so, the specimen is tilted by the maximum angle of the goniometer allowing reaching in a given sense (+Θ). The objective lens is then switched on creating a magnetic field along the optic axis (z-direction) whose component projected in the plane of the thin specimen forces the magnetization to be aligned in a given direction (Figure 5.2(a)). The objective is then switched off, the sample tilted back to zero and the electron holography experiment is carried out to measure the first total phase image: $\phi_1(x) = \phi^E(x) + \phi^M(x)$. The experiment is then repeated but tilting the specimen in the opposite angle (−Θ), therefore, saturating the sample magnetization in the opposite direction (Figure 5.2(b)). The sample is moved back to the zero tilt and the electron holography experiment allows the second total phase shift to be measured: $\phi_2(x) = \phi^E(x) - \phi^M(x)$

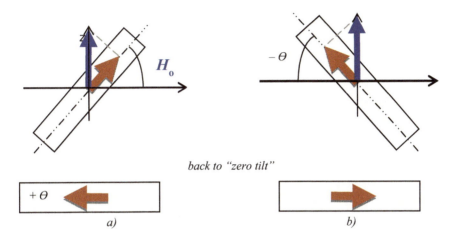

Figure 5.2. *Method to saturate the sample magnetization in two opposite directions using the objective lens magnetic field and the sample tilt*

This method is very convenient when measuring isotropic magnetic materials whose magnetization remains stable in the plane of the thin specimen when switching off the objective lens. It is also works fine in anisotropic materials but, in such cases, the tilt direction has to be controlled in such a way that the magnetization will be aligned along an easy axis when switching on the objective lens.

It, however, does not work for the study of isotropic magnetic nanomaterials whose magnetization is not supposed to be stable in a given direction after saturation. This is the case, for example, for all nanoparticles exhibiting isotropic morphologies (spherical, cubes, stars, etc.). In addition, this method is not suitable for studying a given magnetic configuration as a domain wall separating magnetic domains.

5.2.2.2. *Switching upside down the sample.*

This second method can be used in all cases. It consists of performing a first electron holography experiment on the thin sample to measure the total phase image

$\phi_1(x)$, then removing the specimen from the microscope, switching the sample upside down, reinserting it in the TEM and then recording a second hologram to get the second phase image: $\phi_2(x)$.

The first drawback of this method is that even if the objective lens is switch off, a remnant magnetic field remains far from the optic axis and the specimen is passing through it. This remnant field may modify the magnetic configuration of the specimen leading to a non-perfectly opposite $\phi_1^M(x)$ and $\phi_2^M(x)$ images.

The second issue is the difficulty, when studying dispersed nano-objects, to find back, in a 3 mm TEM grid, a single nano-object of few tens of nanometers large. With patience and determination, this can however be achieved. But, it will be necessary to correct between both the obtained holograms the drift and the rotation introduced during the reversal of the sample.

There is an additional way for separating the $\phi^M(x)$ and $\phi^E(x)$ contributions that consists of taking two electron holograms with two different accelerating voltages. The magnetic phase shift being independent of the electron velocity, only the electrostatic contribution is varying with the electron wave length (see equation [5.3]) and the subtraction of the phase images can provide separately $\phi^M(x)$ and $\phi^E(x)$. This method is, however, not much used, as it implies realigning the microscope at different voltages, which is never straightforward.

5.2.2.3. Image calculation procedure to calculate the magnetic phase image and the electrostatic phase image

We showed that by using the experimental procedure 5.2.2.1 or 5.2.2.2 that $\phi^M(x)$ and $\phi^E(x)$ can be separated following equations [5.4a] and [5.4b]. This, however, necessitates careful image analysis and calculation that are detailed below.

For the extraction of the magnetic phase, two holograms of the same area are needed to be taken together with their corresponding reference holograms. The induction switching is made either by flipping the sample or by applying a magnetic field as detailed in 5.2.2. Amplitude and phase images are then extracted from the two holograms and will be used for the alignment procedure.

In the most general case, three geometrical parameters have to be measured and corrected: (1) the image drift (along the x and y directions in the image), (2) the rotation between them and (3) the magnification changes. The drift is always present as it is impossible to set the sample exactly at the same position in the image within the precision of a single pixel. Rotation and the magnification changes between the two holograms may occur when the sample has been turned upside down. It is, indeed, almost impossible to relocate the area of interest in the exactly same direction with respect to the camera axes when making the reversal process. Even a small rotation of less than a degree requires correction. Finally, even if an identical magnification is set, a variation in scale between two holograms may occur (due to a change in the height of the sample, the focus, etc.). Although these variations are generally small (less than 1% of the magnification), they induce phase shift within the reconstructed images that cannot be neglected.

In the following, we will describe the image reconstruction method for recovering separately the magnetic and electrostatic phase images using two holograms: the "up" part for the first recorded hologram, the "down" part for the hologram acquired after the switching process. After removing numerically the Fresnel fringes due to the biprism, amplitude and phase images are extracted from each hologram (Figure 5.3). The upper images correspond to the "up" configuration whereas the bottom line represents the "down" configuration. Note that the flipping process used to switch the induction induces a large rotation of the region of interest.

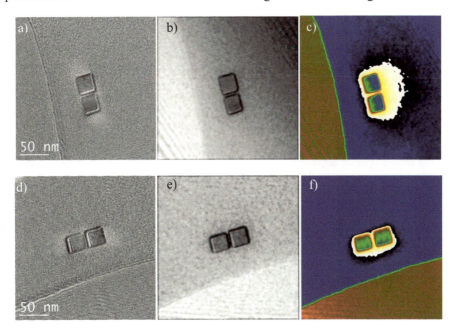

Figure 5.3. *Holograms and corresponding amplitude and phase images obtained from the "up" and "down" experiments: a) "up" hologram, b) "up" amplitude, c) "up" phase, d) "down" hologram, e) "down" amplitude and f) "down" phase*

Then, the "down" images are numerically flipped to take into account the inversion of the sample and, so, the axial symmetry between "up" and "down" parts (Figure 5.4).

Figure 5.4. *Axial symmetry for the "down" amplitude to take into account the reversal process of the sample*

Two common features have then to be found in the images (defects, particular shape, edges,etc.) that will be used to correct all parameters and adjust the two images (drift, rotation and magnification). A line is then drawn manually to connect these features (Figure 5.5): the variation in the line length will give the magnification change between the two images, the relative angle between the two lines will correspond to the rotation, and once these two parameters are corrected, the difference of coordinates of the middle of each line will allow measuring the drift between the "up" and "down" images (both for the amplitude and phase images). When the lines are placed, the user has to choose which of the two images is used as the reference (*from which all the parameters will be calculated*) and which image will be adjusted to the other. For example, the "up" part can be taken as reference images and all parameters for the alignment procedure will be measured and applied to the "down" part.

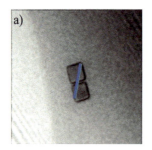

Figure 5.5. *Amplitude images with the lines for the measurement of parameters of alignment: a) "up" amplitude for the reference image, b) "down" amplitude which will be corrected*

The successive corrections are thus the following: the magnification with rotation is adjusted, and then the drift is corrected. The position of the middle of the two lines is recalculated when the first two parameters are corrected. In the case of phase images of weak contrast, it is recommended to use the amplitude images to calculate the corrections to be carried out (Figure 5.6). The best alignment is obtained by subtracting the two "up" and "down" amplitude images that have been realigned and minimizing the resulting contrast. A misalignment will give a strong variation of contrast at the edges. In a final step, the alignment can be slightly improved by applying the alignment procedure directly on the phase images.

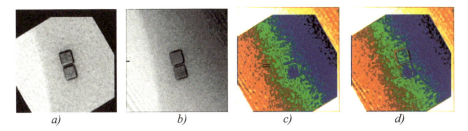

Figure 5.6. *Subtraction between a) the reference "up" amplitude image and b) the corrected "down" amplitude image. The subtraction c) and d) are presented with false color in order to increase the contrast. Subtracted image (c) presents an optimized alignment whereas image (d) corresponds to a misalignment between the amplitude images*

When the superimposition between images has been optimized by adjusting the line positions, we can thus apply the parameters of correction, the half sum and half difference, between the "up" and "down" phase images that give the MIP and the magnetic part, respectively (Figure 5.7). The final phase images have to be renormalized between $-\pi$ and $+\pi$.

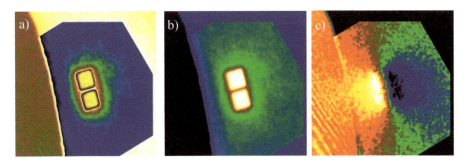

Figure 5.7. *a) MIP phase, b) unwrapped MIP phase and c) magnetic phase, where the magnetic dipole created by Fe nanocubes is clearly visible*

A dedicated software has been developed under Digital Micrograph (Gatan©) to perform these images' corrections and calculations in an intuitive way and with a precision of one pixel (Figure 5.8).

There are, however, other programs of automatic correction that are rather effective, even if the precision is often less accurate, particularly when the magnification and rotation changes are present.

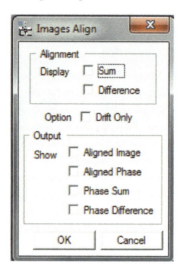

Figure 5.8. *Main window used for the alignment and the choice of output images*

5.3. Hologram analysis: from the phase images to the magnetic properties

Recording in the Lorentz mode, electron holograms on a magnetic sample allow for calculating the phase image of the region of interest:

$$\phi(x) = C_E \int V(x,z) \, dz - \frac{e}{\hbar} \iint B_\perp(x,z) \, dx \, dz = \phi^E(x) + \phi^M(x)$$

We have showed that such a phase image contains both an electrostatic and a magnetic contribution that can be separated. $\phi^M(x)$ can then be obtained and bring quantitative information of the magnetic induction within and in the vicinity of the specimen.

5.3.1. *The simplest case: homogeneous specimen of constant thickness*

If neither V nor $B_\perp$ varies in the incident beam direction, the expression of the phase shift becomes even simpler:

$$\phi(x) = C_E V(x) t(x) - \frac{e}{\hbar} \int B_\perp(x) t(x) dx \qquad [5.7]$$

where $t(x)$ is the sample thickness.

Equation [5.7] can be differentiated with respect to x, leading to:

$$\frac{d\phi^M(x)}{dx} = -\frac{e}{\hbar} B_\perp(x) t(x) \qquad [5.8]$$

$$\frac{d\phi^E(x)}{dx} = C_E \frac{d}{dx}[V(x) t(x)] \qquad [5.9]$$

In the case where the sample is of uniform in-plane thickness and composition (i.e. in x and y directions perpendicular to the incident beam), equation [5.9] goes to zero and the phase gradient of the total phase shift is proportional to the in-plane component of the magnetic induction:

$$\frac{d\phi(x)}{dx} = \frac{d\phi^M(x)}{dx} = -(\frac{e.t}{\hbar}) B_\perp(x)$$

$$B_\perp(x) = -\frac{\hbar}{e.t} \frac{d\phi(x)}{dx} \qquad [5.10]$$

Equation [5.10] indicates that the phase gradient is proportional to the in-plane B component of the induction. Therefore, qualitative mapping of the in-plane magnetic induction can then be simply carried out by plotting the equiphase contours that follow the $B_\perp$ circulation, that is the in-plane magnetic flux.

Quantitative measurements of the magnetic induction can be performed in the vacuum ($t = 0$) and within the specimen in the case when its thickness is determined.

In Figure 5.9 a hologram obtained on two neighboring Fe magnetic nanocubes of 30 nm size (Figure 5.9(a)), the total phase shift $\phi(x,y)$ (Figure 5.9(b)), the electrostatic contribution $\phi^E(x,y)$ (Figure 5.9(c)) and the magnetic contribution $\phi^M(x,y)$ (Figure 5.9(d)) is detailed.

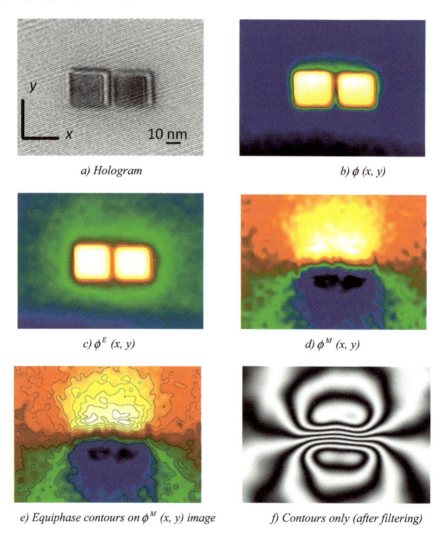

a) Hologram

b) ϕ (x, y)

c) ϕ^E (x, y)

d) ϕ^M (x, y)

e) Equiphase contours on ϕ^M (x, y) image

f) Contours only (after filtering)

Figure 5.9. *Electron holography experiment performed on two Fe neighboring nanocubes*

Equiphase contours have been added over the magnetic phase image (Figure 5.9(e)) and are reported lonely after low pass filtering in Figure 5.9(f). They illustrate how the magnetic flux is spreading out of the two particles that act as a nanomagnet.

A map of the magnetic flux can be drawn by calculating the two components $B_x(x,y)$ and $B_y(x,y)$ of the in-plane magnetic induction. This can be done by extending equation [5.8] in the two dimensions of the magnetic phase image $\phi^M(x,y)$. It goes to:

$$\frac{d}{dx}\phi^M(x,y) = \frac{e.t}{\hbar}B_y(x,y) \qquad [5.11]$$

$$\frac{d}{dy}\phi^M(x,y) = -\frac{e.t}{\hbar}B_x(x,y) \qquad [5.12]$$

and therefore:

$$B_x(x,y) = -\frac{\hbar}{e.t}\frac{d}{dy}\phi^M(x,y)$$

$$B_y(x,y) = \frac{\hbar}{e.t}\frac{d}{dx}\phi^M(x,y)$$

[5.13]

We report in Figure 5.10 the two images of the y and x derivatives of the magnetic phase image, which are proportional to the x and y components of the in-plane magnetic induction, respectively (equation [5.13]).

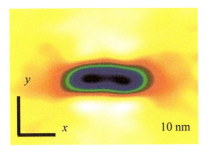
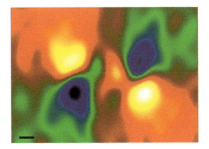

Figure 5.10. *(d / dy) and (d / dx) derivatives of the magnetic phase image) ϕ^M (x, y) proportional to the components B_x (x, y) and B_y (x, y) of the in-plane magnetic induction, respectively*

From such images, one can plot the modulus and the vectorial map of the magnetic induction as shown in Figure 5.11.

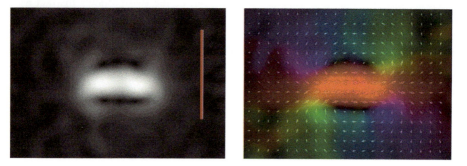

Figure 5.11. *Modulus and vectorial map of the in-plane magnetic flux*

As the thickness of such Fe nanocubes is known from TEM images ($t = 30$ nm), quantitative information on the magnetic properties can be extracted. The most straightforward value is the magnetization M of the nanocube as:

$$\vec{B} = \mu_0(\vec{H} + \vec{M})$$

where H is the applied field (zero in the remnant state, objective lens switch off).

M can then be extracted from equation [5.13]. Figure 5.12 presents a line profile of the integrated magnetic induction extracted from the induction modulus (see Figure 5.5). The maximum value inside the Fe nanocube is equal to 4×10^{-8} T.m. Knowing the thickness of the sample "t", an induction of 1.33 T is measured. *This value is different from the Fe saturation magnetization at 300 K (2.16 T) as the demagnetizing field has to be taken into account for the induction measurement.*

5.3.2. *The general case*

Many specimens do not present so simple a shape and may even be hybrid multicompound systems for, which, the spatial derivatives of the electrostatic phase images do not result in a constant value.

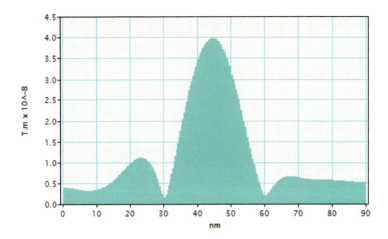

Figure 5.12. *Profile of the magnetic induction across a Fe nanocube (line profile shown in Figure 5.11)*

One of the processes described in section 5.2.2 has to be carried out to separate the electrostatic and the magnetic contributions.

The two-dimensional expression of $\phi^M(x, y)$ in the plane (x, y) perpendicular to the beam direction can be extended from the one-dimensional case of equation [5.5]:

$$\phi^M(x,y) = -\frac{e}{\hbar} \int_{z=-\infty}^{+\infty} \int_{x=\pm\infty}^{x} B_y(x,y,z) \, dxdz \qquad [5.14a]$$

also equal to

$$\phi^M(x,y) = -\frac{e}{\hbar} \int_{z=-\infty}^{+\infty} \int_{y=\pm\infty}^{y} B_x(x,y,z) \, dydz \qquad [5.14b]$$

which allows the in-plane magnetic induction component to be measured.

The magnetic moment along the x and y directions is given by the relations:

$$m_x = \int_{-\infty}^{+\infty} \int_{-\infty}^{+\infty} \int_{-\infty}^{+\infty} B_x(x,y,z) \, dxdydz \qquad [5.15]$$

$$m_y = \int_{-\infty}^{+\infty} \int_{-\infty}^{+\infty} \int_{-\infty}^{+\infty} B_y(x,y,z) \, dxdydz$$

The derivatives of $\phi^M(x,y)$ with respect to x and y gives respectively:

$$\frac{d}{dx}\phi^M(x,y) = -\frac{e}{\hbar}\int_{-\infty}^{+\infty} B_y(x,y,z)dz \qquad [5.16]$$

$$\frac{d}{dy}\phi^M(x,y) = -\frac{e}{\hbar}\int_{-\infty}^{+\infty} B_x(x,y,z)dz \qquad [5.17]$$

The combination of equations [5.14], [5.16] and [5.17] leads to:

$$m_x = -\frac{\hbar}{e}\int_{-\infty}^{+\infty}\int_{-\infty}^{+\infty}\frac{d}{dy}\phi^M(x,y)dxdy \qquad [5.18]$$

$$m_y = -\frac{\hbar}{e}\int_{-\infty}^{+\infty}\int_{-\infty}^{+\infty}\frac{d}{dx}\phi^M(x,y)dxdy \qquad [5.19]$$

The magnetic moment of the specimen can then be calculated by measuring the area under the derivative of the magnetic phase measured in the perpendicular direction following either equation [5.18] or [5.19], choosing the phase image derivative $d/dx\,\phi^M(x,y)$ or $d/dy\,\phi^M(x,y)$, which exhibits the less phase jumps and discontinuities.

5.4. Resolutions

5.4.1. *Magnetic measurements accuracy*

The minimum phase difference between two adjacent pixels the "phase resolution" has been determined by H. Lichte *et al.* [LIC 08]:

$$\Delta\phi_{min} = \frac{SNR}{\mu\sqrt{N_{el}}}\sqrt{2}$$

where μ is the fringes contrast, N_{el} the number of electrons collected per pixel of the camera and SNR the signal-to-noise ratio whose acceptable minimum value is about three.

Contrast: The fringes contrast is decreasing when there is decreasing of the fringes periodicity (i.e. when increasing the voltage on the biprism) but it has to be adjusted in order to have a sufficient number of fringes in the area of interest (see Figures 1.3 and 1.4).

A compromise has then to be made between the number of fringes per nanometers and the fringes contrasts.

The contrast is also increasing when increasing the spatial coherency of the electron beam. We do observe an increase in the fringes contrast when using a cold FEG source (W tungsten tip) compared to a Schottky FEG source whose tip apex is larger.

Electron dose: An electron holography experiment is carried out, typically with a dose of about 500 e⁻/pixel in an FEG microscope. This dose can be smaller but, it therefore, necessitates increasing the exposure time. Increasing the exposure time, however, leads to an increase in the noise associated with mechanical and electrical instabilities of the microscope and possible electromagnetic fields in the area surrounding the microscope. Maximum exposure time in a non-dedicated microscope located in a "good" environment is about 10 s. With an advanced TEM located in a dedicated room, the exposure time may reach 60 s.

We demonstrated that the magnetic phase $\phi^M(x,y)$ obtained by electron holography measures the magnetic flux the electron beam is passing through:

$$\phi^M = \frac{e}{\hbar} \oint Bds$$

with $\Phi_0 = \frac{h}{2e}$, the magnetic flux quantum $\Phi_0 = 2.07 \times 10^{-15}$ Wb (= 2.07 × 10⁻¹⁵ T.m²)

$$\phi^M = \frac{\pi}{\Phi_0} \oint Bds$$

In a non-dedicated microscope, the typical minimum phase shift resolution is about ~$2\pi\ 10^{-3}$ rad. In such a microscope, the minimum magnetic flux an electron holography experiment can measure is about:

$$\Phi_{mag} \sim 2.\ 10^{-3}.\Phi_0 \sim 4.14\ \text{T. nm}^2$$

This corresponds to the magnetic flux spreading out of an iron nanocube of ~2 nm size homogeneously saturated magnetically.

A single Bohr magneton located in a single atom volume (~1 Å^3) would produce a phase shift of ~$2\pi \cdot 10^{-8}$ rad. With a dedicated TEM instrument fitted with a cold FEG source to enhance the spatial coherence of the electron beam, and for which the mechanical vibrations and electrical instabilities have been drastically reduced and located in a calm environment where no additional electromagnetic fields are perturbing and temperature kept constant, we may assume to get a phase resolution 10 times better than the actual solution, that is ~$2\pi \cdot 10^{-4}$ rad, leading to the possibility of detecting about ~10,000 Bohr magnetons.

5.4.2. *Spatial resolution*

The ultimate spatial resolution is given by the resolution of the objective lens used for the experiment. As the objective lens has to be switched off, the spatial resolution in low magnification (LM) mode is very low. In some dedicated instruments, an additional lens called a "Lorentz lens" is located just below the objective lens. It is used in combination with the diffraction lens to create an image of the sample on the detector. Even if such TEM configuration with a dedicated Lorentz lens allows reaching a better spatial resolution than without, these Lorentz lenses have a long focal length (few tens of millimeters) and a large Cs (few meters), which leads to a weak spatial resolution of several nanometers (2–5 nm for the best lenses). Recently, dedicated Cs-corrected instruments that are capable of correcting the spherical aberration of the objective lens and reach spatial resolution of ~50 pm in normal HREM mode, also permit us to correct the spherical aberration of the Lorentz lens, offering an ultimate spatial resolution in Lorentz mode (objective switch off) of ~0.5 nm.

Within this limit, the periodicity of the hologram fringes determines the reconstructable lateral resolution. Therefore, the spatial resolution cannot be separated to the "phase resolution" as a small phase shift can be measured with a high precision but with very poor spatial resolution, that is with large fringes periodicities of high contrast or with poor precision, but high spatial resolution, that is with small fringes periodicities of less contrast.

5.5. One example: FePd (L10) epitaxial thin film exhibiting a perpendicular magnetic anisotropy (PMA)

We report below the electron holography experiment performed on a FePd thin film that exhibits a magnetization oriented perpendicular to the interface. Such magnetic devices can be used for high-density recording media. This magnetic

FePd-L10 order phase has been grown by molecular beam epitaxy at high temperature in order to stabilize with its "c" axis perpendicular to the interface on a FePd disordered phase, grown on a Pd buffer, itself grown on a MgO(001) surface [AUR 09]. The final stacking is FePd$_{L10}$/FePd$_{disord.}$/Pd/MgO(001) as schematized in Figure 5.13.

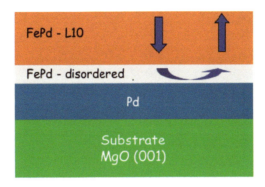

Figure 5.13. *Scheme of the Pd$_{L10}$/FePdd$_{isord.}$/Pd/MgO(001) stacking*

The electrostatic and magnetic phase contributions have been separated switching upside down the sample (method described in section 5.2.2.2). Figure 5.14(a) shows the calculated electrostatic (MIP) contribution to the phase shift and the magnetic contribution is shown in Figure 5.14(b). The isophase contours displayed on both phase images directly relate to thickness variations (in the MIP phase image) and magnetic flux (in the magnetic one).

The variation in the MIP contribution exhibits that the TEM sample increases uniformly in thickness whereas magnetic contribution highlights vortices corresponding to the Bloch walls. Between these vortices are areas where the magnetic flux is parallel or antiparallel to the growth direction. These correspond to the "up" and "down" magnetic domains, that is, the recording bits.

Stray fields close the flux in the vacuum and inside the stack. However, the vortices appear to be flatter at the bottom (close to the FePd$_{disord.}$ layer) than at the top (close to the vacuum). This asymmetrical shape of the vortices is due to the disordered FePd layer that forces the magnetic induction to lie within the foil plane.

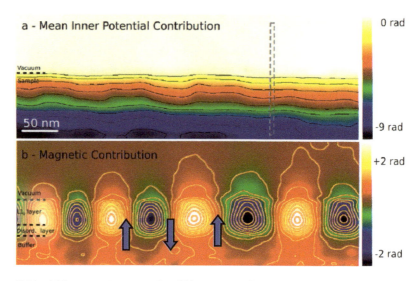

Figure 5.14. *a) Electrostatic potential and b) magnetic phase shift contributions to the phase shift of the electron wave. The equiphase contours represent 1 rad for MIP contribution and 1/4 rad for magnetic contribution*

Quantitative values of magnetic induction can be extracted provided that the MIP or the thicknesses of the different layers are known. Figure 5.15 shows in yellow the experimental profile of the electrostatic contribution to the phase extracted along the dashed line in Figure 5.14. To measure the thickness profile, we have first calculated the MIP values for the Pd, $FePd_{L10}$ and $FePd_{disord.}$ layers.

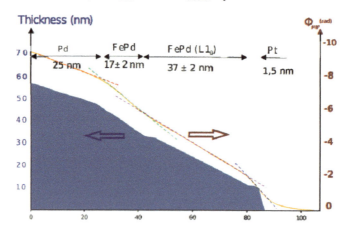

Figure 5.15 *Plot profile along the dashed line of the MIP in Figure 5.6(a). Right (yellow) label and solid line is the phase profile used to deduce the different layers in the foil. Thickness profile is presented in the colored (blue) area and is labeled on the left*

B_x and B_y inside the FePd layer are calculated according to equation [5.13] using the thickness measurements deduced from the MIP contribution and reported in Figures 5.16(a) and (b), respectively. The B_y image clearly shows the magnetic "bits" up and down whereas the B_x image shows how the magnetic flux is closing in the FePd$_{disord.}$ layer.

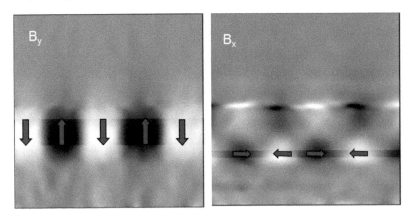

Figure 5.16. *a) B_x and b) B_y components of the magnetic induction deduced from magnetic phase gradients*

Neglecting the demagnetizing field within the material, the measured magnetic induction is directly related to the magnetization in the material. The value of the magnetic induction modulus in the FePd$_{L1_0}$ region (i.e. inside the domains) gives rise to a magnetic induction of 1.3 ± 0.1 T whereas same measurements performed on the FePd$_{disord.}$ area (under domain walls) gives averaged values of 1.2 ± 0.1 T. These $\mu_0 M_s$ values are the same as those expected for bulk FePd. These magnetization measurements do not give the exact magnetic moment but a projection of it in the plane perpendicular to the electron beam.

Accurate measurements of the FePd magnetic properties can, however, be done by performing micromagnetic simulations based on the Landau–Lifshitz–Gilbert equations. Such calculations have been created to simulate the experimental magnetic flux (i.e. the isophase contours) using the magnetic parameters of the bulk materials. Calculated magnetic phase shift and experimental phase images are compared in Figures 5.17(a) and (b).

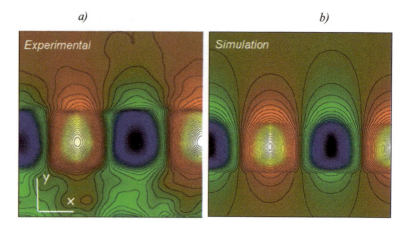

Figures 5.17. *a) and b) Magnified view of Figure 5.7(b) to be compared with a micromagnetic simulation*

We observe that the Bloch walls are much wider in the experimental data. The magnetic flux can be quantitatively measured, both in and outside the sample. The stray field can then be related to the magnetization moment in the domains. This is potentially of great interest for the design of a reader of this kind of material.

5.6. Prospective and new developments

5.6.1. *Enhanced signal and resolution*

We have discussed on the capabilities when using a dedicated microscope fitted with a Cs-corrected Lorentz lens to reach a spatial resolution of few Ångströms and when performing the electron holography experiment on a very stable instrument fitted with highly coherent cold FEG sources and located in a dedicated room to extend the phase resolution up to $\sim 2\pi.10^{-4}$ rad. These are the ultimate spatial and phase resolution achievable now. Further developments are, however, in progress all over the world to push these limits toward atomic resolution such as the "FIRST" project in Japan [FIR].

Other instrumental progresses have also been made in recent years to improve the holograms' signal. For instance, the development of new cold-FEG sources with better spatial coherence allows us to increase the fringes contrast by at least a factor of five [HOU12].

Signal enhancement in holograms can also be obtained by setting a new configuration of biprisms. Fresnel fringes due to the scattering of electrons on the biprism wire can be eliminated by placing a second biprism in the shadow image of the first biprism [HAR 04]. In addition, such an experimental configuration with two biprisms allows, adjusting independently, the fringes periodicity and the overlap area.

We showed that to perform an electron holography experiment, the beam that has interacted with the sample is overlapped with a reference beam that has passed in a vacuum region close to the specimen. Except in case of a perfect flux closure within the magnetic specimen, stray fields are generally spreading around the sample and then are perturbing the reference beam passing close to it. Recently, T. Tanigaki showed that when using a biprism in the condenser system, holographic interferences can be obtained from the overlap of two regions that can be far away from each other avoiding the stray fields perturbation of the reference beam [TAN 12]. *This method will also be useful for electrostatic measurements.*

5.6.2. *In-situ switching*

We have described the objective lens high magnetic field issue when measuring magnetic properties of specimen in a TEM and the way of performing electron holography experiments in Lorentz mode in order to keep the sample in a field-free environment, that is in its remnant state.

It is, however, interesting to study the magnetic specimen response to the application of a controlled magnetic field. This can be achieved by getting a dedicated magnetizing stage that consists of a sample holder on which two coils have been added on each side of the specimen to apply a magnetic field that can typically reach ~200 Oe [BUD 11].

Another method is to tune the current sent in the objective lens to control the magnetic field applied to the specimen [RAF 02, JAV 10]. Because the magnetic field within the pole piece is parallel to the optic axis (z-direction), the sample has to be tilted in order to adjust the projection of the field in the sample plane (see Figure 5.2). When controlling the two α and β sample tilts and the current sent in the coils, the objective allows tuning the components of the field projected in the sample plane and the field can be reached few thousand Oe. The drawback of this method of applying a magnetic field is that the applied field cannot be strictly in a single direction of the sample as a strong component perpendicular to it always exists. It, however, can provide some information on the magnetic switching and domain configuration changes with the applied field for samples with a high shape anisotropy (such as thin films).

In Figure 5.18 the local study by electron holography of the magnetic switching of a tunneling magnetic junction (TMR) used in magnetic reading heads is reported

Au/Co/Fe/MgO/FeV/MgO(001). The magnetic phase images have been obtained at the continuous values of the sample tilt at a constant value of the applied magnetic field. They show, locally, the two magnetic layers {Co/Fe} and {FeV}, independent switching and the whole TMR passing from a parallel to antiparallel configuration.

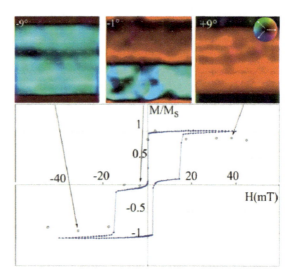

Figure 5.18. *Magnetic phase images obtained on a Co/Fe/MgO/FeV stacking for three values of the tilt, that is of the applied magnetic field superimposed of the hysteresis loop obtained by VSM (Images are 50 nm width)*

Whatever the chosen method, the main issue when studying a specimen in TEM under the application of an external magnetic field is that the external field is also tilting the electron beam. The resulting distortion of the image can be quite important at high field and difficult to correct.

5.7. Conclusions

The rising of spintronic and magnetic devices in the microelectronic industry requires the development of advanced experimental methods for the local measurements of their structural, chemical and magnetic properties. We have demonstrated that electron holography is a highly powerful technique that allows not only the local measurement of the specimen magnetization and the fine analysis of the magnetic domain configurations, but it is also a suitable tool for the study of magnetic switching processes in these devices. Even if this experimental technique may not appear straightforward for common TEM users, we believe that, as for strain measurements performed in dark-field electron holography, it will become a key method for the study of forthcoming advanced magnetic devices.

5.8. Bibliography

[AHA 59] AHARANOV Y., BOHM D., "Significance of electromagnetic potentials in quantum theory", *Phys Rev*, vol. 115, p. 485, 1959.

[AUR 09] MASSEBOEUF A., MARTY A., BAYLE-GUILLEMAUD P., GATEL C. and SNOECK, E. "Quantitative observation of magnetic flux distribution in new magnetic films for future high density recording media", *NanoLett.*, vol. 9, p. 2803, 2009.

[BEL 03] BELEGGIA, M., SCHOFIELD M.A., ZHU Y., MALAC M., LIU Z. and FREEMAN M. "Quantitative comparison of magnetic field mapping in TEM with micromagnetic simulations", *Appl. Phys. Lett.*, vol. 83, p. 1435, 2003.

[BUD 11] BUDRUK A., PHATAK C., PETFORD-LONG A.K. and DE GRAEF M., "In-situ Lorentz TEM magnetization studies on a Fe-Pd-Co martensitic alloy", *ActaMaterialia*, vol. 59, p. 6646, 2011.

[CHA 07] CHAPPERT C., FERT A., VAN DAU F.N., "The emergence of spin electronics in data storage", *Nature Mat.*, vol. 6, pp. 813–823, 2007.

[DUN 98] DUNIN BORKOWSKI R.E., MCCARTNEY M.R., SMITH D.J. and PARKIN S.S.P., "Towards quantitative electron holography of magnetic thin films using in situ magnetization reversal", *Ultramicroscopy*, vol. 74, p. 61, 1998.

[DUN 00] DUNIN BORKOWSKI R.E., MCCARTNEY M.R., KARDYNAL B., PARKIN S. S. P., and SMITH D.J., Off-axis electron holography of patterned magnetic nanostructures. *J. Microscopy*, vol. 200, p. 187, 2000.

[DUN 02] DUNIN-BORKOWSKI R.E. and MCCARTNEY M.R., Chapter in *Advances in nanophase materials and nanotechnology, Volume: Magnetic Nanostructures*, in NALWA H.S. (ed.), American Scientific Publishers, pp. 299–325, 2002.

[FIR] First, http://www.first-tonomura-pj.net/e/index.html.

[HAR 04] HARADA K., TONOMURA A., TOGAWA Y., AKASHI T. and MATSUDA T., "Double-biprism electron interferometry", *Appl. Phys. Lett.*, vol. 84, p. 3229, 2004.

[HOU 12] HOUDELLIER F., MASSEBOEUF A., MONTHIOUX M., HŸTCH M.J., "New carbon cone nanotip for use in a highly coherent cold field emission electron microscope", *Carbon*, vol. 50, pp. 2037–2044, 2012.

[HŸT 11] HŸTCH M.J., HOUDELLIER F., HÜE F., SNOECK E., "Dark-field electron holography for the measurement of geometric phase", *Ultramicroscopy*, vol. 111, no. 8, p. 1328, 2011.

[JAV 10] JAVON E., MASSEBOEUF A., GATEL C. and SNOECK E., "Electron holography study of the local magnetic switching process in magnetic tunnel junctions", *J. Appl. Phys.* vol. 107, 09D310, 2010.

[LIC 08] LICHTE H., "Performance limits of electron holography", *Ultramicroscopy* vol. 108, pp. 256–262, 2008.

[MCC 97] MCCARTNEY M.R. and SMITH D.J., "Electron Holography and Lorentz Microscopy of Magnetic thin Films and Multilayers", *Scanning Microscopy*, vol. 11, p. 335, 1997.

[MCC 98] MCCARTNEY M.R. and ZHU Y., Off-axis electron holographic mapping of magnetic domains in Nd2Fe14B. *J. Appl. Phys.* vol. 83, p. 6414, 1998.

[SIC 11] SICKMANN J., FORMÁNEK P., LINCK M., MUEHLE U., LICHTE H., "Imaging modes for potential mapping in semiconductor devices by electron holography with improved lateral resolution", *Ultramicroscopy*, vol. 111, pp. 290–302, 2011.

[SNO 03] SNOECK E., DUNIN-BORKOWSKI R.E., DUMESTRE F., RENAUD P., AMIENS, C. CHAUDRET B. and ZUCHER P., "Quantitative magnetization measurements on nanometer ferromagnetic cobalt wires using electron holography", *Appl. Phys. Lett.*, vol. 82, p. 88, 2003.

[SNO 08] SNOECK E., GATEL C., LACROIX L.M., BLON T., LACHAIZE S., CARREY J., "RESPAUD M., CHAUDRET B., Magnetic configurations of 30 nm iron nanocubes studied by electron holography", *Nano Lett.*, vol. 8, no. 12, p. 4293, 2008.

[TAN 12] TANIGAKI T., INADA Y., AIZAWA S., SUZUKI T., PARK H.S., MATSUDA T., TANIYAMA A., SHINDO D. and TONOMURA A., "Split-illumination electron holography", *Appl. Phys. Lett.*, vol. 101, p. 043101, 2012.

[TON 92] TONOMURA A., "Electron-holographic interference microscopy", *Adv. Phys.*, vol. 41, p. 59, 1992.

Chapter 6

Interdiffusion and Chemical Reaction at Interfaces by TEM/EELS

6.1. Introduction

The purpose of this chapter is to show how research on interdiffusion and chemical reactions in thin films and their interface with semiconductor substrates can be conducted using the transmission electron microscopy (TEM) based method and their coupling with electron energy-loss spectroscopy (EELS). Analytical TEMs commonly installed nowadays in most materials science laboratories are mainly considered in this study. The discussion will be illustrated in the context of the developments of new gate oxides (commonly called high-k or HK) for the replacement of SiO_2 in metal oxide semiconductor field-effect transistors (MOSFETs). Literature data from HfO_2, which is already integrated in MOSFET production since end of 2007 [BOH 07] and of rare earth-based oxides (REO), will be particularly discussed together with results obtained at CEMES.

6.2. Importance of interfaces in MOSFETs

For several decades, the microelectronics industry succeeded in reducing the size of the components with an outstanding efficiency and regularity. The gate length of the key component, the MOSFETs, has scaled down from several microns in 1971 to some tens of nanometers nowadays. The Si/SiO_2 couple, the heart of microelectronics devices, was the chosen ally for the successful scaling because it is

---

Chapter written by Sylvie SCHAMM-CHARDON.

a system that presents several intrinsic qualities unrivaled to date. First, an amorphous SiO_2 layer can be thermally grown on silicon with excellent control in thickness and uniformity, and forms a natural very stable interface with the silicon substrate, with a low density of intrinsic interface defects. Second, SiO_2 shows an excellent thermal stability in contact with silicon, a property that is required for the fabrication of advanced MOSFETs, which includes thermal process steps up to 1,000°C. Third, SiO_2 has a suitable band alignment with respect to silicon, which is needed to reduce the tunneling current flowing through the gate insulator. Finally, the compatibility of SiO_2 with poly-Si gate electrodes in a self-aligned complimentary MOS technology is also a determining factor in their successful integration into advanced devices.

However, the golden age of silica reached the end in the early 21st Century. In fact, the International Technology Roadmap for Semiconductors (ITRS) specifications imposed the SiO_2 thickness to a scale below 1 nm in order to improve the performances of the components [ITR 03]. This was problematic because of the large leakage of currents flowing through the MOS structure by a quantum mechanical tunneling mechanism. After 10 years of intense research [WIL 01, HOU 06, ROB 06, WON 06], the need to reduce the gate leakage current and static power consumption in Si-based complementary MOSFETs led to the replacement of the SiO_2 gate dielectric by OXIDE layers with higher dielectric constants (HK) and the replacement of poly-Si as the gate electrode by metal gates (MG). There is still intense research activity on HfO_2-based dielectrics but other oxides are also being investigated, often involving rare earths (RE) [FAN 06]. At that time, several problems had to be faced. Among them, the thermal processing required to make a device caused both physical and chemical changes in the system including regrowth of SiO_2 and/or the formation of silicate-like interfacial layer (IL) at the Si/HK interface. While decreasing the effective dielectric constant, if done in a controlled manner, such an IL could be beneficial as it separates the carriers in the channel from the trapped charge and defects in the HK dielectric and so stops degradation of the channel mobility. The high thermal budgets also resulted in crystallization, phase separation, diffusion across interfaces and significant interfacial reactions, particularly at the Si/HK and HK/MG boundaries. Such changes could alter the effective work function and have a serious impact on the device performance. The nature and extent of such interactions was dependent on the deposition techniques used as well as the details of the processing and the thermal budget required. Therefore, development of these stacks was imposed to control their stability along the fabrication process. Control of interfaces and nanovolumes imposed to rely on characterization methods dedicated to studies at the nanoscale level, such as nano-analytical TEM, a key method increasingly used to perform an investigation of the matter at this level.

6.3. TEM and EELS

The power of a nano-analytical TEM refers to its ability to give access to the topographical and structural investigation of a sample using different imaging modes in a TEM experiment and to combine this information with the chemical composition accessible via EELS performed in the same experiment. The structural, chemical and even electronic properties of materials can nowadays be rather routinely investigated on a nanometric-scale with analytical TEMs commonly installed in most materials science laboratories and it is the purpose of this chapter to give some practical information and demonstration of the way it can be performed.

We will mainly consider the case of modern TEMs working at 200–300 keV and equipped with a field emission gun (FEG), a post-column annular dark field (ADF) or high-angle annular dark field (HAADF) detector and an electron energy-loss spectrometer. This spectrometer is usually positioned post-column but in-column spectrometer (Omega) also exists. CCD cameras are classically used for image formation.

For the purpose of atomic structure imaging and analysis with a spatial resolution at least in the nanometer range, a FEG source of electrons is necessary in order to get a stable, coherent, optically bright and small electron probe. (HA)ADF is a necessary support for local EELS analysis in TEM. It is used to locate the area of interest of the analysis, as will be illustrated later. The EELS spectrometer selects electrons of the source as a function of the energy they have lost by interacting with the electrons of the samples. As a consequence, a chemical selectivity is accessible and all the phases present can be analyzed, amorphous as well as crystalline. Note that the use of a corrector for spherical aberration C_s is of particular importance if we want to study interfaces and what happens nearby with the high-resolution mode of the TEM (HRTEM). In fact, the image C_s correction is dedicated for the direct observation of atomic structures at interfaces with substantially reduced contrast delocalization in the images at interfaces and surfaces [HAI 98].

6.4. TEM/EELS and study of interdiffusion/chemical reaction at interfaces in microelectronics

TEM associated with EELS can address some important points in relation with the many requirements for alternative gate dielectrics for Si-based MOSFETs. Many studies have already been devoted to HK materials by the community of microscopists with different equipments and different techniques, but all aim to solve the same problems related to HK dielectric layers. They are as follows.

Precise thickness determination of the gate oxide film and of the unavoidably formed IL. In fact, precision is important because even a 0.1 nm decrease in the oxide thickness can lead to an order of magnitude increase in leakage current [TIM 98].

Crystallization state in the gate dielectric film before and after post-deposition annealing treatments. Whether it is amorphous, polycrystalline or epitaxial, it has noticeable consequences. Polycrystalline films, which present grain boundaries and interfacial roughness, are believed to be correlated to an undesired increase in leakage currents and reduction in transistor mobility [KIN 00] but a stable epitaxial structure could also avoid a low permittivity SiO_2-rich IL formation. If amorphous materials were first preferred by the semiconductor industry, the subnanometer equivalent oxide thickness objective will probably not be reached without crystalline HK.

Quality of the interface. In fact, the knowledge of the roughness of the atomic arrangement at the interface between Si and the HK will give an insight into the associated electronic structure in order to understand the electrical properties.

Chemical nature of the different coexisting phases in the stack. This point must be systematically addressed together with the previous three. For example, the calculation of an important parameter that characterizes the strength of the HK, that is the equivalent oxide thickness of the film as defined in [WIL 01] (EOT: theoretical thickness of SiO_2 that would be required to achieve the same capacitance density as the dielectric = physical thickness of the HK scaled by the ratio of its dielectric constant to that of SiO_2) can be done if the dielectric properties of the layers are known (chemical nature known) together with their physical thickness.

In the following sections, examples from the literature will be used to illustrate how practically TEM-EELS can address these different points.

6.4.1. *Thickness measurement*

Precise measurement of the oxide thickness is critical. Many examples of the dielectric and the IL thickness analysis are found in the literature. They support EOT determinations [STE 01, TRI 04], model for X-ray reflectivity (XRR) data fitting [GUP 01] and capacitance measurements or estimation of the dielectric constants of amorphous phases [MER 04].

High-resolution lattice images are a reliable tool for precise thickness determination. The images can be obtained either with a coherent, phase-contrast imaging, that is with HRTEM or with an incoherent, amplitude-contrast technique, that is HAADF imaging. For the first case, sample thickness (the thickness traveled by the electron beam of the electron microscope) must be thin enough (10–20 nm) to avoid some contrast artifact in the image of the interface of the film with the substrate due to roughness of this interface. For the second case, thickness can be measured in sample as thick as 50 nm provided a large inner angle detector and a highly localized probe are used [MUL 01, DIE 03]. A comparative and detailed investigation of dielectric film thickness determination by HRTEM and HAADF can be found in [DIE 03]. In general, thickness measured by HRTEM can be 0.5 nm thinner than the one determined by HAADF-STEM [MUL 01, DU 04]. The interface roughness is the major source of error in a ultrathin gate dielectric thickness measurement. Averaging of the roughness occurs over the sample thickness (traveled by the electrons). Both oxide thickness (1.5 nm) and interface roughness (0.175 and 0.3 nm) have been measured with HAADF images for the stack c-Si/SiO_2/Poly-Si [MUL 01].

6.4.2. *Atomic structure analysis*

In order to study the atomic structure of an interface, the TEM sample is prepared as a cross-section so that when observed in the TEM, it is oriented in such a way that the electron beam is parallel to the film/substrate interface, and to a particular zone axis of the substrate. The electron microscopy modes are similar to those previously described for thickness measurements.

HRTEM is a parallel detection method in which a nearly parallel 200–300 keV electron beam simultaneously illuminates the entire imaged area of the crystal. The primary and elastically scattered beams interfere and form the image by a coherent superposition. The image is related to the projected potential of the object (atomic positions) and therefore "represents" the crystalline structure of the specimen for specific intervals of thickness and defocus. At 200 keV, for Si and a Cs correction of 1 µm, it has been shown that the Si dumbbells are typically observed for a defocus close to 0 and several nanometers thick lamella [CAS 10]. This last condition can easily be obtained with TEM samples prepared classically (see section 6.4.4). Images of thicker samples cannot be interpreted directly and need to be compared to computer simulations in order to be able to understand the more complicated contrasts (contrast reversals for example). This is in general the case if samples are prepared with a focused ion beam (FIB, 80–100 nm), for example when a MOSFET has to be located on a large wafer. Hopefully, recent developments of FIB at low voltages allow going down to 20- to 40-nm thickness samples [SCH 12].

HAADF is a serial detection method where a sub-0.25 nm diameter probe (atomic dimensions) of 100–300 keV electrons is focused and scanned across the imaged area of the crystal. On microscopes equipped with a HAADF detector, large angle elastic scattering can be detected and incoherently summed to obtain an imaging signal that depends strongly on the atomic number of the atoms illuminated by the probe (Z-contrast image). Moreover, because the nominal probe size is smaller than many interatomic distances, the scattered signal reveals atomic columns with good contrast [BAT 93]. The thicker the sample, the better is the contrast with an upper limit of approximately 50 nm. Because the image is acquired serially pixel-by-pixel, the atomic position cannot be determined as accurately as with a parallel detection method. Because of its sensitivity to Z-contrast, an ADF image is also a first approach of chemical composition. This is particularly useful in the case of HK oxides deposited on Si often with SiO_2-based IL. In ADF images, the bright contrast corresponds to the HK and dark contrast to Si and SiO_2, SiO_2 appearing darker than Si. Therefore, in the case of HK, HAADF is often used for both chemical investigation and precise probe positioning during EELS analysis and for atomic structure imaging [DU 04, LYS 06] (see section 6.5).

Lattice imaging is used in the context of HK thin films developments to check the crystallization state of the films and also to follow the chemical stability of the interface between the Si substrate and the HK during the different steps of the process fabrication, the annealing step being the most important one. Because the possibility to use Cs corrected microscopes, it is now comfortable to locate interfaces and to "read" atomic structures at these interfaces (see section 6.5). Interfaces can appear sharp, with a direct transition between the (001) planes of Si substrate and the amorphous or crystalline phase of the HK observed with a dark contrast on HREM images (or a white contrast in a HAADF image). But often an IL of bright contrast resulting from a reaction at the interface and/or an interdiffusion is seen on the HRTEM images (dark contrast in HAADF images). Sometimes, for amorphous HK films, there is no particular change in the contrast of the image near the interface and it is difficult to conclude about the existence of a reaction [STE 01, TRI 05]. Moreover, it is seen in the literature that the same HK/Si stacks prepared in different ways can lead to interaction or no interaction depending on the deposition method (physical vapor deposition or chemical vapor deposition), and parameters. This point was discussed for REO-based HK such as La_2O_3 [TRI 05], Gd_2O_3 [GUP 01] and Pr_2O_3 [ONO 01] cases. For these reasons, it is of great importance to go further in the investigation by using analytical techniques. X-ray photoelectron spectroscopy (XPS) is most systematically used [GUP 01, TRI 05, EDG 04, LO 05, HEN 12, CHI 12] but medium-energy ion scattering spectroscopy (MEIS) [MAR 01, GUH 00], Fourier transform infrared spectroscopy (FT-IR) [ONO 01, KWO 01] or secondary ion mass spectroscopy (SIMS) [TRI 08] have also been investigated. Literature results of HfO_2, which is already integrated in MOSFET production since the end of 2007 [BOH 07] and of REO will be used to illustrate the different modes of TEM-EELS investigations.

6.4.3. EELS analysis

EELS has the advantages of being performed in line with the TEM experiment. It adds a new dimension to the structural imaging with the electron microscope that is a high-resolution chemical analysis or imaging. The information provided by EELS results from the electron–specimen interactions. These are elastic and inelastic scattering processes. The incident electrons have a high energy (typically between 100 and 300 keV). The basic signal is the EELS spectrum, which represents the scattered intensity as a function of the decrease in kinetic energy of the fast electrons. The EELS spectrum of La_2O_3 is shown as an example in Figure 6.1.

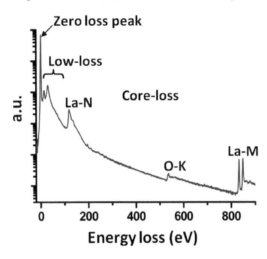

Figure 6.1. *Electron energy-loss spectrum of a reference La_2O_3 film shown with the electron intensity on a logarithmic scale to evidence the large dynamic range of intensities going from the low-loss region (plasmon) to the core-loss region (ionization edges of each element)*

We find there the zero-loss or "elastic" peak and, in the low energy-loss domain, the main signature is the plasmon peak. The zero-loss peak represents electrons that are transmitted without suffering any measurable energy loss. The plasmon peak corresponds to the collective excitation of valence or conduction electrons. Then, edges are seen in the core-loss energy domain, above ~100 eV, superimposed on a decreasing background, nearly as a power law of energy. They indicate the individual excitations of inner-shell electrons. The corresponding sharp rise in intensity occurs at the ionization threshold whose energy is approximately equal to the binding energy of the corresponding atomic shell. From the detailed study of the EELS spectrum, information about specimen thickness, dielectric response, gap (low energy-loss domain), elemental composition, chemical bonding and band structure (core-loss energy domain) is accessible, with an energy resolution depending essentially on the source of the TEM electron beam and a spatial resolution limited by the TEM optics and the uncertainty principle [EGE 11]. Depending on the apparatus on which it is performed, EELS can be used in different ways. Conventional TEM-EELS probes elemental composition and electronic structure with energy resolution and spatial resolution that depend on the electron emission. Typical values are 0.7 eV or 1 eV and 1 nm or 10 nm in the case of a Schottky field emission or a thermionic emission, respectively. 0.3 eV energy resolution and subnanometer spatial resolution is obtained with cold FEG. Better performances, 1 Å and 0.1 eV, are nowadays accessible with probe corrected [BAT 02] and monochromated [MIT 03] electron sources, respectively. Energy-filtered TEM (EFTEM) maps elemental composition down to 1-nm spatial resolution. These two modes have been performed for the study of HfO_2 and REO thin films on Si.

The classical way to perform EELS is to locate the electron probe at places of interest and acquire the EELS spectrum. For example, HRTEM and EELS performed on as-deposited and annealed Gd and La silicate films allow us to see the modification of the topography of the interfaces with annealing temperature together with the crystallization state of the film and the chemical nature of the IL [WU 02]. This method can be automated by scanning the probe over the area of interest, that is over a line or a surface. It was initially introduced as the spectrum imaging method in the particular case of scanning TEMs [JEA 89]. This method is described in section 6.5 in the case of a conventional TEM equipped with a scanning stage.

Another technique largely used for studying HK thin films is the combination of HAADF imaging and EELS performed in the scanning TEM mode (HAADF STEM-EELS). In this case, field emission sources of electrons allowing probe sizes as small as 0.2 nm are used. A lattice resolution Z-contrast image is formed to position the small electron probe at chosen locations within the layers. Often line scans are performed across the interface from the Si substrate to the gate dielectric. A pretty and invaluable example of this technique has been proposed in [MUL 99]. The electronic structure at the atomic scale of ultrathin gate oxides is studied there. It is shown by comparing the O-K edges of interfacial and bulk SiO_2, that for a 1-nm oxide thickness (five silicon atoms across), the fundamental thickness limit of a usable SiO_2 gate dielectric is 0.7 nm. Features in the fine structures (ELNES) of the O-K edge are used for this experiment, particularly the modification of the O-K edge at the Si/SiO_2 interface is compared with the one of bulk SiO_2. HAADF STEM-EELS has also been used to illustrate La diffusion into a thermal oxide about 2 nm thick after annealing at 800°C in N_2 ambient. The demonstration was carried out by looking at the decreasing intensity of the La-$M45$ edge as the STEM probe moves closer to the interface [STE 01]. The HfO_2 case has also been studied this way in order to understand the effect of nitrogen incorporation in the gate stack [LYS 06] and to discriminate about the possible formation of a silicate-like compound at the HfO_2/Si interface [AUG 95].

Recording a spectrum map that contains a series of EELS spectra acquired simultaneously as a function of a spatial coordinate, for example the direction perpendicular to the substrate/film interface, is another approach of TEM-EELS. This method has been called with different terms: spatially resolved EELS [REI 95, KIM 99], laterally resolved EELS [GOL 03] or EELS profiling [WAL 03]. Contrary to STEM, TEM-EELS is intrinsically free from energy shifts that can occur when spectra are acquired sequentially with a small focused probe. Moreover, beam damage can be reduced because a wide lateral area is analyzed. Despite these advantages, this is practically not as easy as to use as STEM-based techniques. This approach can be performed with a TEM equipped with an imaging filter (2D CCD detection) operated in the spectroscopy mode, for which a line focus perpendicular to the energy dispersion direction is used (line focus EFTEM). An important and delicate point is that the film/substrate interface must be precisely aligned with the energy-dispersive direction of the filter. In order to overcome this difficulty, a rotation holder can be used [BOT 02].

Instead of probing a focused electron beam in HAADF STEM-EELS or of recording a spectrum map in line focus EFTEM, bidimensional chemical maps with a high spatial resolution can be recorded with a TEM equipped with an energy filter operated in the imaging filtered mode [REI 95]. This image mode allows us to separate the contributions from elastically and inelastically scattered electrons by inserting an energy-selecting slit in the energy-dispersive plane of a filter. Then, zero-loss filtering, plasmon filtering or ionization edge filtering is possible. For the case of ionization edge filtering, the contribution of the background is eliminated by recording several filtered images around the edge (referred as the two-window or three-window methods). For Pr_2O_3 films, EFTEM of O-K, Pr-M and Si-L maps have been recorded using the three-window method and the corresponding line profile have been extracted across the Pr_2O_3 film and the Si substrate. By comparing the filtered images, a two-layer interface is identified as already confirmed by HRTEM images [LO 03].

6.4.4. *Sample preparation*

One essential point for the microscopists is the quality of the TEM sample. This means getting a thin enough sample for atomic structure imaging with a rather constant thickness for EELS investigations and with minimized preparation artifacts. The TEM-EELS experiment will give the best quality results if the sample preparation is good enough.

Generally, TEM works proposed on HK/Si stacks have been performed on samples prepared by standard procedures (case (3) in the following list). Let us cite one example among the few works where different methods were tested on a same stack. No differences were found between the observations made on a Si/3.5 nm HfO_2/1 nm HfSiO/200 nm poly-Si stack prepared using (1) FIB, (2) cross-sectional tripod polishing or (3) standard grinding, polishing, dimpling and ion milling [MAC 03]. On the contrary, in [DU 04], conventional mechanical polishing (case (3)) with minimal ion milling (less than 5 min) and plasma cleaning is considered as the best way to obtain satisfactory results for ultrathin gate dielectric chemical and structural studies. The FIB thinning followed by lift-out with C supporting grid (case (1)) was found to introduce detrimental contamination effects on EELS and HAADF analysis results and also to induce major deterioration of the high-resolution HAADF-STEM images. The study was performed on the Si/bare oxide/3 nm HfO2/200 nm poly-Si gate stack. Note that for the particular case of REO, it is necessary to be careful with the use of water during the polishing treatment because of the reactivity and even the solubility of REO in water, as mentioned in the case of thin Gd_2O_3 layers [KWO 01].

As already mentioned, the best situation for good TEM-EELS interpretations is to have a thin enough sample, in the range of 20–50 nm. In this range, the thinnest samples will be well dedicated for atomic structure imaging and the thickest for EELS investigations.

The following section will illustrate, in a concrete way, how TEM/EELS has been used to contribute to the development of a next generation of HK thin films, based on RE and transition metal (TM) elements with the aim of proposing alternatives for the currently used HfO_2.

6.5. HRTEM/EELS as a support to developments of RE- and TM-based HK thin films on Si and Ge

6.5.1. *Introduction*

Downscaling in microelectronics raises big challenges particularly concerning the choice of new materials and their integration in metal-oxide-semiconductor (MOS) structures for logic applications [ROB 08]. Research is now focused on gate oxides with high permittivity (k) and semiconductor substrates with high mobility. A lot of effort is particularly devoted to the engineering of HK/semiconductor stacks with the lowest EOT. In these stacks, two main contributions have to be considered. First, an IL of low dielectric constant unavoidably forms during deposition and transforms due to postannealing treatments that is detrimental for the total EOT of the stack. To limit this effect, it is important to minimize the thickness of this IL and to maximize its k value. Second, the HK thin film just above the IL should have a higher k value than HfO_2, stable even for low physical thicknesses (Figure 6.2).

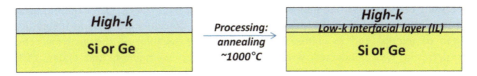

Figure 6.2. *With processing, an unavoidable low-k IL is formed between the HK and the semiconductor film*

To reach these objectives, the binary oxide La_2O_3 and the ternary compound ZrO_2 doped with La were considered. In fact, La_2O_3 is expected to crystallize above 400°C in the hexagonal phase (h-La_2O_3), which has a higher k value ($k \sim 27$) than the cubic phase (c-La_2O_3) in which most of REOs crystallize. Moreover, doping of TM oxides have been reported to stabilize metastable phases, cubic and tetragonal ones, with higher dielectric constant ($k \sim 30$–40) than their monoclinic polymorph. However, the level of reactivity of these oxides with Si (or Ge) and their ability to form an IL had to be elucidated. Our purpose was to provide a way to describe at the nanometer level the structural and chemical states of each layer constituting the stack and also to try to estimate their dielectric contribution. The atomic structure and the elemental profiles across the stacks were investigated at the nanometric level and the thickness of each layer constituting the stack measured on the basis of HRTEM/EELS experiments. In parallel, the dielectric constant of the stack was determined by capacitance–voltage measurements. By introducing these structural, chemical and dielectric parameters in a multiple-layers capacitor model describing the succession of the layers in the stack, the dielectric constant of each layer was estimated and the values obtained discussed taking into account the elemental composition and the stabilized crystallographic phases. The results of the measurements have revealed atomic diffusion and reactions during deposition and postdeposition annealing that will be described in the following.

6.5.2. HRTEM/EELS methodology

6.5.2.1. The microscope/TEM sample

For these investigations, the microscope used was a Schottky field emission TEM, FEI Tecnai™ F20 microscope operating at 200 kV. The microscope is equipped with a corrector for spherical aberration of the objective lens (CEOS). As already said, this is a prerequisite for the direct observation of atomic structures at interfaces with contrast delocalization minimized [HAI 98]. For local EELS studies, the microscope is also equipped with a scanning stage allowing a focused 1-nm-sized probe to be scanned over the sample area of interest (in our case, a line crossing the film/substrate interface) and an imaging filter (Gatan GIF Tridiem) used as a spectrometer (2k × 2k sensor). ADF/HAADF detectors are used to image the area of interest and to choose on the image the location of the analysis. The experiment was conducted with the FEI, Digital Micrograph and Gatan user interfaces.

Specimens transparent to electrons were prepared for cross-sectional (XTEM, bulk is cut perpendicular to the HK film/substrate interface) and plan-view (PVTEM, bulk is cut parallel to the HK film/substrate interface) observations using the standard procedure: mechanical standard grinding, polishing and minimal Ar^+ ion milling.

6.5.2.2. *Experimental methodology*

After controlling the homogeneity of the films by performing electron diffraction pattern analyses and dark field imaging over micrometric areas on PVTEM, XTEM samples are observed at medium magnification, ×145,000. With the 2k × 2k CCD camera installed on the microscope, this enables to appreciate the homogeneity of the film thickness over a hundred nanometers with atomic resolution kept thus allowing an absolute calibration of the image using the known projected parameters of Si, which is oriented [110].

Homogeneity being confirmed, local analyses can be conducted. HRTEM images (typically ×490,000) with a rather parallel electron beam are acquired on different areas and the crystallization state within the stack, the thickness of the layers constituting the stack and eventual roughness are determined. An example of HRTEM image where these parameters are evidenced is given in Figure 6.3. An amorphous IL with uniform contrast appears between the Si[110] projected structure and the planes of the polycrystalline HK. Its thickness and the thickness of the HK layer can be measured with an estimated accuracy of one Si (001) atomic plane, that is ±0.5 nm. Hereafter, we will call the thicknesses deduced from HRTEM images, structural thickness t_S. HRTEM images are obtained from rather thin areas (few nanometers) in order to optimize the signal to background ratio and to limit the effect of the projected roughness on the determination of the frontiers of the different layers of the stack. At the Si/HK film interface, the roughness was limited to one or two (001) atomic planes whereas on Ge the roughness was larger.

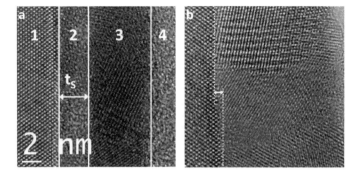

Figure 6.3. *HRTEM images where a) different layers of the stack can be distinguished by their crystallization state (1: Si[110] projected structure; 2: amorphous IL with a t_S thickness; 3: polycrystalline layer; 4: amorphous glue used for the TEM sample preparation); white lines define the frontiers of the layers and b) roughness at a Ge substrate surface with a step of 3 (001) atomic planes*

For the EELS analysis, an area of interest is imaged by HRTEM (Figure 6.4(a)). Then a focused 1-nm sized beam is formed that will define the spatial resolution of the EELS analysis and will be used for the ADF/HAADF image acquisition (Figure 6.4(b)). The ADF/HAADF image is calibrated with respect to the HRTEM image and is used to set up a line across the Si/HK layer stack (dashed line on Figure 6.4(b)) that will be scanned by the 1-nm probe for the acquisition of EELS spectra. The step between two acquisition points on the line is chosen the same as the probe size, that is 1 nm, in order to avoid overlap of analyzed zones and the line is positioned with a 30° angle with respect to the HK layer/semiconductor substrate in order to have a final effective step of 0.5 nm. At each focused point of the line, EELS spectra encompassing ionization edges of the elements of interest (Si, Ge, O, La, Zr) are collected. In fact, according to the use of the 2k × 2k camera of the Gatan GIF Tridiem and the choice of a dispersion of 0.3 eV per channel, four energy windows of 600 eV range are necessary to cover all the available edges (see Tables 6.1 and 6.2). However, the spectra of the first energy window corresponding to the shortest acquisition time is the only one considered for quantification because it corresponds to a high enough signal-to-noise ratio for reasonable quantification (minimal intensity at the edge of the elements higher than 1,000 counts) and its acquisition induces a very limited damage of the sample compared to the larger duration over few to several seconds (see Chapter 2 for radiation damage discussion). The other spectra are only used for cross-correlation to support interpretation of the results. The low energy loss spectra are also acquired to determine the relative thickness of the area analyzed. The reduced thickness must be low enough, that is $t/\lambda < 0.5$, in order to neglect the effect of multiple scattering. Figure 6.5 represents an example of a spectrum line reconstructed from the different energy windows including the low-loss region. After the quantification of the spectra, which will be detailed below, the relative compositions of the elements are plotted as a function of the distance from the semiconductor surface (Figure 6.9(b)). The spatial calibration takes into account the thermal drift of the sample. This drift is estimated from ADF images obtained just before and just after the spectrum line acquisition.

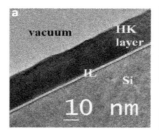
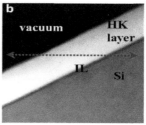

Figure 6.4. *a) HRTEM (x145000) and b) ADF image of the area of interest. IL appears with a bright (dark) contrast in the HRTEM (ADF) image. The blue dotted line defines the locations where the 1 nm probe is scanned to obtain the EELS spectra*

Tension	Probe size	Collection angle	Convergence angle	Energy dispersion	Energy resolution	Reduced thickness[1]
200 keV	1 nm	20 mrad	7 mrad	0.3 eV/channel	1.2 eV	0.2–0.5

Table 6.1. *Experimental parameters for the EELS spectra acquisition on La3O2 or LaZrO/Si or Ge stacks*

Energy window for acquisition (eV)	[50–650]	[450–1,050]	[700–1,300]	[1,750–2,350]
Acquisition duration (s)	0.5	3	5	10
Si-L (99 eV)[2]	X			
La-N (99 eV)[2]	X			
Zr-M (180 eV)[2]	X			
O-K (532 eV)[2]	X	X		
La-M (832 eV)		X	X	
Ge-L (1217 eV)[2]			X	
Si-K (1839 eV)				X
Zr-L (2222 eV)				X

Table 6.2. *Experimental parameters for the EELS spectra acquisition on La2O3 or LaZrO/Si or Ge stacks*

---

1 Reduced thickness is the ratio of the physical thickness to the inelastic mean free path t/λ (at 200 keV : λ is approximately 100 nm).
2 Ionization edges considered for the quantitative analysis.

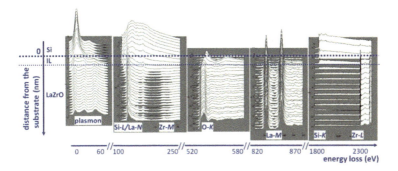

Figure 6.5. *Reconstructed spectrum line from four different energy windows (see Table 6.2) including the low-energy loss domain acquired along a linear trajectory going from a Si substrate toward a LaZrO-based HK layer. For a better reading of the ionization edges, they have been extracted from their background and their intensities are magnified*

6.5.2.3. *Stray scattering and Ge detection*

The generation of secondary and backscattered electrons in commercial Schottky FEG is a problem for EELS investigations [MCC 97, SCO 01]. Some of them can pass through the anode and be accelerated and appears in the EELS spectrum as added features, which may confuse spectral analysis. The position of the extra features depends on the extraction and gun lens parameters used for the emission of the electrons. For the conditions we used (EXT 3,800 eV and GL5), the maximum of the feature appeared approximately 1,155 eV, in a region placed just before the Ge-L edge. Another added smoother contribution to the EELS spectrum was noticed in the 450 eV region (O-K region). These added contributions have to be removed before any quantification of the EELS spectra. For this, we simply acquired for each spectrum line a spectrum in the vacuum and subtracted this spectrum from the spectrum line. The result seems convincing as shown on Figure 6.6.

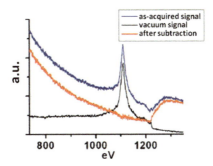

Figure 6.6. *Subtraction of the EELS signal in the vacuum (black) from the experimental signal on the stack where Ge is present (blue) in order to recover a Ge-L edge ready for quantification (red)*

6.5.2.4. Details on the quantification procedure

In general, classical quantitative analysis is applied to ionization edges except when two edges are superimposed in the EELS spectrum. This was the case for the Si-L and the La-N edges located at the same 99-eV energy position.

For the classical quantification, the ratio of the areal density of each element with respect to O was determined with the following relation:

$$\frac{N_A}{N_{O-K}} = \frac{I_A(\beta,\Delta)}{I_{O-K}(\beta,\Delta)} \cdot \frac{\sigma_{O-K}(\beta,\Delta)}{\sigma_A(\beta,\Delta)}$$

where $I(\beta,\Delta)$ is the integrated intensity below the ionization edge over the energy window Δ after extraction of the background and $\sigma(\beta,\Delta)$ is the calculated partial ionization cross-section corresponding to the core level analyzed [EGE 11] (Figure 6.7). β is the collection angle. A represents the Si-L, La-N or Zr-M edges.

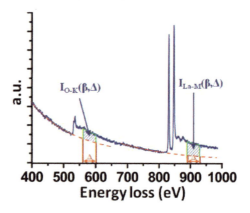

Figure 6.7. *Quantitative analysis of an ionization edge: background extraction (red), calculation of the integrated intensity below the ionization edge over the energy window Δ (hatched area)*

The result of this analysis is dependent on the way the background is extracted below the ionization edges, on the positioning and width of the integration energy window, and also on the accuracy of the calculated cross-section, which is limited for of L, M and N edges [EGE 11]. Therefore, in order to minimize these uncertainties, reference samples of known composition were first analyzed with the same parameters for acquisition and treatment. From this analysis, correction factors named currently k-factors are determined as:

$$k = \frac{\left(\dfrac{N_A}{N_{O-K}}\right)_{nominal}}{\left(\dfrac{N_A}{N_{O-K}}\right)_{experimental}}$$

The k-factors for our study are summarized in Table 6.3. They are used in order to be able to propose quantified elemental profiles.

Reference sample	Ionization edge	First energy window	Width of the energy window	k-factor (O-K)
SiO_2	Si-L	99 eV	40 eV	2.00 ± 0.10^3
h-La_2O_3	La-N	99 eV	40 eV	0.36 ± 0.05^3
m-ZrO_2	Zr-M	180 eV	40 eV	2.20 ± 0.15^3

Table 6.3. *Experimentally determined k-factors*

When edges are superimposed like Si-L and La-N edges at positions nearby the interface of the HK/Si stack, the multiple linear least square (MMLS) method [LEA 91] can be used. With this method, the experimental spectrum is simulated with a linear combination of reference spectra using a least square approach. The method is easily accessible because it is implemented on the Digital Micrograph software. The reference spectra that can be used are the one of Si-L in Si and in SiO2 (no reference spectra exists for silicate compounds of La) and the La-N edge in La2O3. From our experience, it is preferable to use spectra from the same spectrum line where the Si-L/La-N edge to be simulated was acquired. The Si-L edge obtained from the Si substrate and the La-N edge obtained on the La2O3 or LaZrO-based HK layer away from the interface were found relevant references. It was not always possible to obtain a Si-L reference of SiO$_2$ in the same spectrum line. In this case, a reference edge from a silica layer acquired in the same conditions was used. Finally, from the MMLS simulation performed over the 95–150 eV energy range, the proportion of Si in the Si compounds and of La in the La compounds are extracted. The final intensities of the deduced Si-L and La-N profiles are calibrated with the ones obtained by classical quantification in the Si substrate and in the HK La$_2$O$_3$ or LaZrO layer, respectively

---

3 Values represent the dispersion observed over the different measures.

6.5.2.5. Positioning the interfaces: structural and chemical thicknesses

Looking at the elemental profiles of Figure 6.9(b), it is clear that three regions corresponding to different elemental compositions can be defined. Two regions of constant composition, that is in the Si substrate and in the HK layer away from the interface, are surrounding a region of variable composition. This last region defines the IL chemical extent. A question remains: where to place the frontiers between these three regions? This positioning will depend on the nature of the interface, abrupt or not, and also on the probe size and the probe effective step used to obtain the profiles. Let us consider the case of our experiment. For this, it is useful to simulate the profiles of a model stack like $Si/SiO_2/(La_2O_3)_x(SiO_2)_y$ (y/x linearly increasing/decreasing)/La_2O_3 where each layer is 2 nm thick. With a probe size of 1 nm and a step of 0.5 nm, the elemental area profiles for this stack were simulated taking into account a geometrical contribution of each element where the probe covers two phases. For example, it means that a probe centered at the Si/SiO_2 interface will give an intensity for half a part proportional to the one of Si in the pure Si phase and for the other half part an intensity proportional to the one of Si in pure SiO_2. These intensities must integrate the atomic densities of the different phases, that is 5.02×10^{22} at/cm^3 for Si and 6.60×10^{22} at/cm^3 for SiO_2 (note for La_2O_3, density is 5.42×10^{22} at/cm^3). Following this approach, the simulated profiles for the model stack are shown in Figure 6.8 and compared to the geometric position of the interfaces. The inflexion point on the Si or on the O profiles defines the position of the abrupt Si/SiO_2 interface whereas for the two other interfaces defined between a constant composition domain (SiO_2 or La_2O_3) and a variable composition domain (($La_2O_3)_x(SiO_2)_y$), it is the change of slope that defines the position of the interfaces. If we transpose this approach to the real case of Figure 6.9(b), we are able to define the Si/IL frontier at the position of the inflexion point of the O profile and the IL/HK layer frontier when the Si intensity profile goes to 0. The distance between these two chemical frontiers will be defined as the chemical IL thickness, t_C. This chemical thickness is always different from the structural thickness as determined from the HRTEM image. It is larger by a few nanometers. A roughness at the IL/HK layer where there is a mixing of the polycrystalline HK with an amorphous silica-rich phase could probably account at least for a part for this difference.

6.5.2.6. Determination of the dielectric constant of the different layers within the stack

The chemical and structural thicknesses determinations of the different layers from HRTEM and EELS were particularly useful for the determination of the dielectric constant of the different layers of the stack according to the following procedure.

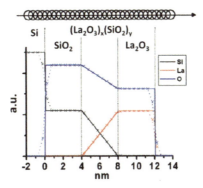

Figure 6.8. *Geometrical position of the different layers in the Si/SiO2/(La2O3)x(SiO2)y/La2O3 stack (full lines) and simulated chemical profiles (crosses). Probe, 1 nm; step, 0.5 nm*

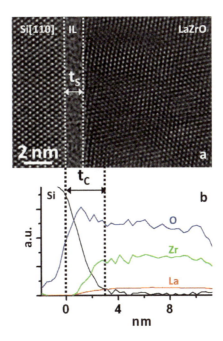

Figure 6.9. *a) HRTEM image of a Si/LaZrO-based HK layer and b) the corresponding EELS elemental profiles obtained from a spectrum line like the one of Figure 6.4. Structural and chemical thicknesses, respectively, t_S and t_C, are indicated on the HRTEM image and elemental profiles*

If we rely on the case of Figure 6.9, we can describe the stack as a two-layer stack comprising the low-k IL and the HK oxide. This stack can be thus modeled by two capacitors in series with an EOT written as follows:

$$EOT = \frac{3.9}{k_{low-k}} t_{low-k} + \frac{3.9}{k_{high-k}} t_{high-k}$$

From this relation, the dielectric constant of both the low-k IL and the HK oxide can be determined provided the physical thickness of these layers and the EOT of the corresponding stack are known at least for two different stack thicknesses. HRTEM and EELS can afford information about the layer thicknesses whereas EOT can be measured from MIS capacitors. This has been done systematically for different stacks [SCH 09, SCH 11] and will be rapidly illustrated in the following section.

6.5.3. Illustrations

The following cases illustrate the development of atomic layer deposited (ALD) films on Si and Ge and the optimization of the ALD parameters in order to obtain stacks with the lowest EOT stabilizing the HK oxide with the highest k polymorph and minimizing the low-k IL layer thickness or maximizing its k value.

6.5.3.1. La_2O_3 /Si

Thin REO films were considered among the candidate high dielectric constant materials to substitute HfO_2 as the next-generation gate dielectric in advanced microelectronic devices. Among the REO series, the first member La_2O_3 is the most promising oxide because, stabilized in the hexagonal phase (h-La_2O_3) above 400°C, it features the highest k value ($k = 27$) [SCA 07, TSO 08]. This prediction motivates the effort to produce h-La_2O_3 in thin layers grown by ALD. However, it is also known that La_2O_3 is the most hygroscopic oxide of the REO series exhibiting the highest affinity for Si atoms. This was confirmed by our investigations that showed a high reactivity of La_2O_3 with Si forming an amorphous La-silicate IL even in as-grown films (Figure 6.10) [SCH 09, LAM 10].

The use of O_3 instead of H_2O as an oxygen precursor of the ALD process improves the quality of the stack, reducing the thickness of the IL ($k = 11$) and stabilizing the HK hexagonal phase of La_2O_3 for thinner film thicknesses (100 nm against 20 nm) [LAM 10]. These studies lead to the identification of a "critical thickness" that, although reduced, still limits an aggressive scaling down of the

EOT. But, using hexagonal La$_2$O$_3$ in combination with a second HK oxide could result in a promising solution.

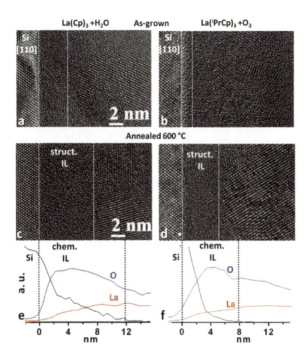

Figure 6.10. *(a–d) HRTEM images and (e, f) EELS elemental profiles of ALD grown La$_2$O$_3$-based films on Si with different precursors. Evidence of an amorphous La-silicate IL formation at the interface, the thickness of which increases dramatically with annealing (compare (a) to (c) or (b) to (d))*

6.5.3.2. La$_x$Zr$_{1-x}$O$_{2-\delta}$ (x = 0.25)/Si

ZrO$_2$ could be another option for replacing HfO$_2$ because it satisfies technology requirements in terms of k value and band gap energy but the material is marginally chemically/thermally stable on Si, thus restricting an aggressive scaling down of the equivalent oxide thickness. Alloying ZrO$_2$ with La$_2$O$_3$ to produce a ternary La$_x$Zr$_{1-x}$O$_{2-\delta}$ compound might be a promising route to engineer a material suitable for HK dielectric applications. Improvement with respect to La$_2$O$_3$-based stacks was clearly demonstrated. Even an amorphous silicate IL is formed, its extent is much reduced and the annealing treatment increases its thickness smoothly. Note that the EELS analysis is able to evidence a phase separation in the IL upon annealing that cannot be suspected by HRTEM imaging. La and Zr elements are absent over 1 nm just above the interface for the annealed film contrary to the case of the as-grown film (Figure 6.11). Phase separation is more pronounced for thicker HK films. On the one side, this induces a detrimental decrease of the k value from 7.5 to 5.5 of the IL due to silica enrichment, but on the other side, it is associated to a favorable decrease of the density of interface traps from 3.8×10^{12} eV^{-1} cm^{-2}. It was also demonstrated that annealing induces a significant increase of the k value of the HK film from 17.5 to 30 [TSO 09]. This can be explained by the large increase of the crystallization fraction of the cubic polymorph of ZrO_2 as seen on the electron diffraction pattern and the dark-field images in Figure 6.11.

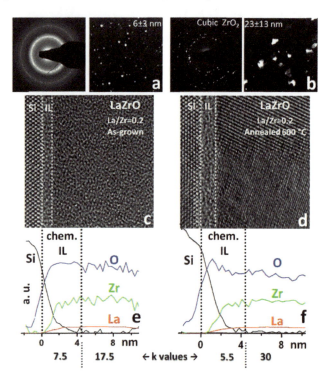

Figure 6.11. *Selected area electron diffraction patterns (SAED), dark-field images (DF), HRTEM images and EELS elemental profiles of ALD-grown $La_xZr_{1-x}O_{2-\delta}$ films on Si, as-deposited (left) and annealed (right). Evidence of an amorphous silicate IL formation at the interface, which phase separates in the annealed stack, being silica-rich just above the interface. (a,b) SAED and DF modifications upon annealing are consistent with the increase of the k value observed (e, f)*

6.5.3.3. *La-doped ZrO_2/Ge*

As an alternative to Si in high-speed logic devices, Ge is widely considered due to its higher carrier mobilities and, among HK oxides, ZrO_2 has been proved to be a promising insulator. However, interface engineering still remains an open issue as the Ge/oxide interface exhibits an unacceptably large density of interface traps. On the basis of works that proved RE doping being an efficient way of passivating interface defects, ALD-grown La-doped ZrO2 films were deposited on Ge substrates and their structural and chemical properties examined. Contrary to the Si case, there is no structural IL formation with Ge (Figure 6.12). However, a chemical IL is evidenced by the EELS elemental profiles where a La peak of concentration is observed just above the interface over 2 nm regardless of the annealing process. The main difference between the as-grown and annealed films is the Ge out-diffusion observed after annealing. It is proved that Ge atoms are supplied by the substrate and penetrate into the oxide upon annealing. The significant increase of the HK k value after annealing is consistent with Ge-induced stabilization of the tetragonal ZrO_2 phase, which has the highest k value and the inhibition of the lower k monoclinic ZrO_2 polymorph. This is proved by the examination of the electron diffraction pattern of the HK film. Ge diffusion occurs, reasonably affecting the roughness at the interface qualified by a La-based germanate-like region, which is accompanied by an acceptably low density of interface traps [SCH 11, LAM 09].

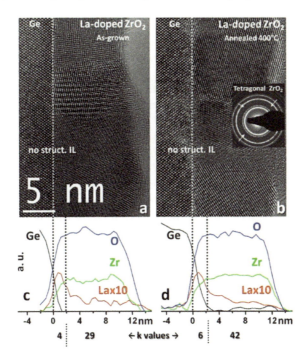

Figure 6.12. *HRTEM images and EELS elemental profiles of ALD grown La-doped ZrO_2 films on Ge as-deposited (left) and annealed (right). No evidence of a structural IL (a, b) but clear evidence of a chemical IL (c, d) and of Ge out-diffusion after annealing (d). SAED confirms the stabilization of the tetragonal phase of ZrO_2*

6.6. Conclusion

Processing of nanometer-sized devices causes physical and chemical changes that had to be identified for an efficient control of their properties. A TEM coupled with EELS has increasingly been investigated for this purpose providing nano-analytical studies at the required nanoscale level. In particular, HRTEM can currently afford atomic structure images of the corresponding MIS stacks with a precise and rather direct reading of the atomic position at interfaces providing the microscope used is equipped with a correction for spherical aberration of the objective lens and the sample studied is thin enough. The corresponding commercial microscopes were installed over the last years in most materials science laboratories. EELS spectrometer (filter) are now also installed on these microscopes. Their use provides an added chemical dimension to the investigations, which is shown to be essential for an accurate insight in the organization of materials even if it is limited to the nanometer level as demonstrated in this chapter. Moreover, one must know that rather recently, instrumentation in TEMs has reached an unrivaled spatial resolution at the ångström level, thus allowing the acquisition of maps, atomic column by atomic column, of the nature of the atoms across interfaces. Even the bonding state is accessible due to now attainable ~0.1 eV energy resolutions [MUL 08, COL 10, COL 11, KRI 11]. The significant development in computer assistance, both in hardware and in software, has fully contributed to these spectacular improvements. However, one must not forget that TEM-EELS progress is also dependent on the quality of the electron transparent sample. For investigations of micro-nanoelectronic real devices, this became tricky because of the need of a site-specific method of preparation. FIB sample preparation is becoming the most relevant and efficient technique for this purpose [SCH 12] (see Chapter 9). When a interface is of a concern, roughness has also to be taken into consideration in an accurate manner [INA 10, CRA 12]. Because current devices imply elements with greatly varying mass density, the EELS collection angle has to be considered with care [BOS 10]. Finally, in the context of ever increasing probe currents, radiation damage is more of a concern (see Chapter 2). It is particularly pronounced for low Z-elements [EGE 04]. Low accelerating voltage experiments are proposed today to mitigate this difficulty [COL 11, BOT 10, SUE 10a, SUE 10b, SUE09, SUE10].

6.7. Bibliography

[AUG 06] AUGUSTIN M.P., BERSUKER G., FORAN B., BOATNER L.A., STEMMER S, "Scanning transmission electron microscopy investigations of interfacial layers in HfO2 gate stacks", *Journal of Applied Physics*, vol. 100, 024103, 2006.

[BAT 93] BATSON P.E., "Simultaneous STEM imaging and electron energy-loss spectroscopy with atomic-column sensitivity", *Letters to Nature*, vol. 366, pp. 727–728, 1993.

[BAT 02] BATSON P., DELBY N., KRIVANEK O.L., "Electron energy-loss near-edge structures of 3d transition metal oxides recorded at high-energy resolution", *Nature*, vol. 418, pp. 617–620, 2002.

[BOH 07] BOHR M.T., CHAU R.S., GHANI T., MISTRY K., "The High-k Solution", *IEEE Spectrum*, vol. 44, pp. 29–35, 2007.

[BOS 10] BOSMAN M., ZHANG Y., CHENG C.K., LI X., WU X., PEY K.L., LIN C.T., CHEN Y.W., HSU S.H., HSU C.H., "The distribution of chemical elements in Al- or La-capped high-k metal gate stack", *Applied Physics Letters*, vol. 97, 103504, 2010.

[BOT 02] BOTTON G.A., GUPTA J.A., LANDHEER D., MCCAFFREY J.P., SPROULE G.I., GRAHAM M.J., "Electron energy loss spectroscopy of interfacial layer formation in Gd_2O_3 films deposited directly on Si (001)", *Journal of Applied Physics*, vol. 91, pp. 2921–2928, 2002.

[BOT 10] BOTTON G.A., LAZAR S., DWYER C., "Elemental mapping at the atomic scale using low accelerating voltages", *Ultramicroscopy*, vol. 110, pp. 926–934, 2010.

[CAS 10] CASANOVE M.J., COMBE N., HOUDELLIER F., HŸTCH M.J., "Visualising alloy fluctuations by spherical-aberration–corrected HRTEM", *Europhysics Letters*, vol. 91, 36001, 2010.

[CHI 12] CHIANG C.-K., WU C.-H., LIU C.-C., LIN J.-F., YANG C.-L., WU J.-Y., WANG S.-J. "Characterization of $Hf_{1-x}Zr_xO_2$ Gate Dielectrics with $0 < x < 1$ Prepared by atomic layer deposition for metal oxide semiconductor field effect transistor applications", *Japanese Journal of Applied Physics*, vol. 51, 011101, 2012.

[COL 10] COLLIEX C., BOCHER L., DE LA PENA F., GLOTER A., MARCH K., WALLS M., "Atomic-scale STEM-EELS mapping across functional interfaces", *Journal of Materials*, vol. 62, pp. 53–57, 2010.

[COL 11] COLLIEX C., "From electron energy-loss spectroscopy to multi-dimensional and multi-signal electron microscopy", *Journal of Electron Microscopy*, vol. 60, suppl. 1, pp. S161–S171, 2011.

[CRA 12] CRAVEN A.J., SCHAFFER B., SARAHAN M.C., "Analysing interface reactions using EELS", *Journal of Physics: Conference Series*, vol. 371, 012008, 2012.

[DIE 03] DIEBOLD A.C., FORAN B., KISIELOWSKY C., MULLER D.A., PENNYCOOK S.J., PRINCIPE E., STEMMER S., "Thin dielectric film thickness determination by advanced transmission electron microscopy", *Microscopy and Microanalysis*, vol. 9, pp. 493–508, 2003.

[DU 04] DU A.Y., TUNG C.H., FREITAG B.H., ZHANG W.Y., LIM S., ANG E.H., NG D., "Ultra-thin SiON and high-k HfO_2 gate dielectric metrology using transmission electron microscopy", *Proceedings Of the 11th International Symposium on the Physical and Failure Analysis of Integrated Circuits*, pp. 135–138, 2004.

[EDG 04] EDGE L.F., SCHLOM D.G., BREWER R.T., CHABAL Y.J., WILLIAMS J.R., CHAMBERS S.A., HINKLE C., LUCOVSKY G., YANG Y., STEMMER S., COPEL M., HOLLÄNDER B., SCHUBERT J., "Suppression of subcutaneous oxidation during the deposition of amorphous lanthanum aluminate on silicon", *Applied Physics Letters*, vol. 84, pp. 4629–4631, 2004.

[EGE 04] EGERTON R.F., LI P., MALAC M., "Radiation damage in the TEM and SEM", *Micron*, vol. 35, pp. 399–409, 2004.

[EGE 11] EGERTON R.F., *Electron Energy-Loss Spectroscopy in the Electron Microscope*, 3rd. ed., Springer Science + Business Media, LLC, New York, 2011.

[FAN 06] FANCIULLI M., SCAREL G. (EDS), "Rare earth oxide thin films", *Topics in Applied Physics*, vol. 106, Springer-Verlag, 2006.

[GOL 03] GOLLA-SCHINDLER U., BENNER G., PUTNIS A., "Laterally resolved EELS for ELNES mapping of the Fe L_{23}- and O K-edge", *Ultramicroscopy*, vol. 96, P. *573*, 2003.

[GUH 00] GUHA S., CARTIER E., GRIBELYUK M.A., GRIBELYUK N.A., BOJARCZUK N.A., COPEL M.C., "Atomic beam deposition of lanthanum- and yttrium-based oxide thin films for gate dielectrics", *Applied Physics Letters*, vol. 77, pp. 2710–2712, 2000.

[GUP 01] GUPTA J.A., LANDHEER D., SPROULE G.I., MCCAFFREY J.P., GRAHAM M.J., YANG K.-C., LU Z.-H., LENNARD W.N., "Interfacial layer formation in Gd_2O_3 films deposited directly on Si (001)", *Applied Surface Science*, vol. 173, pp. 318–326, 2001.

[HAI 98] HAIDER M., ROSE H., UHLEMANN S., SCHWAN E., KABIUS B., URBAN K., "A spherical aberration-corrected 200 kV transmission electron microscope", *Ultramicroscopy*, vol. 75, no. 1, pp. 53–60, 1998.

[HEN 12] HENKEL C., HELLSTRÖM P.-E., ÖSTLING M., STÖGER-POLLACH M., BETHGE O., BERTAGNOLLI E., "Impact of oxidation and reduction annealing on the electrical properties of $Ge/La_2O_3/ZrO_2$ gate stacks", *Solid State Electronics*, vol. 74, pp. 7–12, 2012.

[HOU 06] HOUSSA M., PANTISANO L., RAGNARSSON L.A., DEGRAEVE R., SCHRAM T., POUTOIS G., DE GENDT S., GROESENEKEN G., HEYNS M.M., "Electrical properties of high-k gate dielectrics: challenges, current issues, and possible solutions", *Material Science and Engineering: Research*, vol. 51, pp. 37–85, 2006.

[INA 10] INAMOTO S., YAMASAKI J., OKUNISHI E., KAKUSHIMA K., IWAI H., TANAKA N., "Annealing effects on a high-k lanthanum oxide film on Si (001) analyzed by aberration-corrected transmission electron microscopy/scanning transmission electron microscopy and electron energy loss spectroscopy", *Journal of Applied Physics*, vol. 107, 124510, 2010.

[ITR 03] ITRS 2003, available at http://www.itrs.net/Links/2003ITRS/Home2003.htm.

[JEA 89] JEANGUILLAUME C., COLLIEX C., "Spectrum-image: The next step in EELS digital acquisition and processing", *Ultramicroscopy*, vol. 28, p. 252–257, 1989.

[KIN 00] KINGON A., MARIA J.P., STREIFFER S.K., "Spectrum-image: the next step in EELS digital acquisition and processing", *Nature*, vol. 406, pp. 1032–1038, 2000.

[KIM 99] KIMOTO K., KOBAYASHI K., AOYAMA T., MITSUI Y., "Analyses of composition and chemical shift of silicon oxynitride film using energy-filtering transmission electron microscope based spatially resolved electron energy loss spectroscopy", *Micron*, vol. 30, pp. 121–127, 1999.

[KWO 01] KWO J., HONG M., KORTAN A.R., "Properties of high kappa gate dielectrics Gd_2O_3 and Y_2O_3 for Si", *Journal of Applied Physics*, vol. 89, pp. 3920–3927, 2001.

[KRI 11] KRIVANEK O.L., DELLBY N., MURFITT M.F., "Aberration-corrected scanning transmission electron microscopy of semiconductors", *Journal of Physics: Conference Series*, vol. 326, 012005, 2011.

[LAM 09] LAMAGNA L., WIEMER C., BALDOVINO S., MOLLE A., PEREGO M., SCHAMM-CHARDON S., COULON P.E., FANCIULLI M., "Thermally induced permittivity enhancement in La-doped ZrO2 grown by atomic layer deposition on Ge(100)", *Applied Physics Letters*, vol. 95, p. 122902, 2009.

[LAM 10] LAMAGNA L., WIEMER C., PEREGO M., VOLKOS S.N., BALDOVINO S., TSOUTSOU D., SCHAMM-CHARDON S., COULON P.E., FANCIULLI M., "O_3-based atomic layer deposition of hexagonal La_2O_3 films on Si(100) and Ge(100) substrates", *Journal of Applied Physics*, vol. 108, p. 084108, 2010.

[LEA 91] LEAPMAN R.D., HUNT J.A., "Comparison of detection limits for EELS and EDXS", *Microscopy Microanalysis Microstructures*, vol. 2, pp. 231–244, 1991.

[LO 03] LO NIGRO R., TORO R.G., MALANDRINO G., RAINERI V., FRAGALA I.L., "A simple route to the synthesis of Pr_2O_3 high-k thin films", *Advanced Materials*, vol. 15, pp. 1071–1075, 2003.

[LO 05] LO NIGRO R., TORO R.G., MALANDRINO G., CONDORELLI G.G., RAINERI V., FRAGALA I.L., "Praseodymium Silicate as a high-k dielectric candidate: An insight into the Pr2O3-film/Si-Substrate interface fabricated through a metal-organic chemical vapor deposition process", *Advanced Functional Materials*, vol. 15, pp. 838–845, 2005.

[LYS 06] LYSAGHT P.S., BERSUKER G., TSENG H.-H., JAMMY R., "Spectroscopic analysis of the process-dependent microstructure of ultra-thin high-k gate dielectric film systems", *Surface Interface Analysis*, vol. 38, pp. 1588–1593, 2006.

[MAC 03] MACKENZIE M., CRAVEN A.J., MCCOMB D.W., HAMILTON D.A., MCFADZEAN S., "Spectrum imaging of high-k dielectric stacks", *Institute of Physics Conference Series*, vol. 199, p. 299, 2003.

[MAR 01] MARIA J.P., WICAKSANA D., KINGON A.I., BUSCH B., SCHULTE H., GARFUNKEL E., GUSTAFSSON T., "High temperature stability in lanthanum and zirconia-based gate dielectrics", *Journal of Applied Physics*, vol. 90, pp. 3476–3482, 2001.

[MCC 97] MCCOMB D.W., WEATHERLY G.C., "The effect of secondary electrons generated in a commercial FEG-TEM on electron energy-loss spectra", *Ultramicroscopy*, vol. 68, pp. 61–67, 1997.

[MER 04] MEREU B., DIMOULAS A., VELLIANITIS G., APOSTOLOPOULOS G., SCHOLZ R., ALEXE M., "Interface trap density in amorphous $La_2Hf_2O_7$ /SiO_2 high-k gate stacks on Si", *Applied Physics A*, vol. 80, p. 253, 2004.

[MIT 03] MITTERBAUER C., KOTHLEITNER G., GROGGER W., ZANDBERGEN H., FREITAG B., TIEMEIJER P., HOFER F., "Electron energy-loss near-edge structures of 3d transition metal oxides recorded at high-energy resolution", *Ultramicroscopy*, vol. 96, pp. 469–480, 2003.

[MUL 99] MULLER D.A., SORSCH T., MOCCIO S., BAUMANN F.H., EVANS-LUTTERODT K., TIMP G., "The electronic structure at the atomic scale of ultrathin gate dielectrics", *Nature*, vol. 399, pp. 758–761, 1999.

[MUL 01] MULLER D.A., "Gate dielectric metrology using advanced TEM measurements", *AIP-Conference-Proceedings*, vol. 550, pp. 500–505, 2001.

[MUL 08] MULLER D.A., FITTING KOURKOUTIS L., MURFITT M., SONG J.H., HWANG H.Y., SILCOX J., DELLBY N., KRIVANEK O.L., "Atomic-scale chemical imaging of composition and bonding by aberration-corrected microscopy", *Science*, vol. 319, pp. 1073–1076, 2008.

[ONO 01] ONO H., KATSUMATA T., "Interfacial reactions between thin rare-earth-metal oxide films and Si substrates", *Applied Physics Letters*, vol. 78, pp. 1832–1834, 2001.

[REI 95] REIMER L. (Ed), *Energy-Filtering Transmission Electron Microscopy*, Springer, Berlin-Heidelberg-New-York, 1995.

[ROB 06] ROBERTSON J., "High dielectric constant gate oxides for metal oxide Si Transistors", *Rep. Prog. Phys*, vol. 69, pp. 327–396, 2006.

[ROB 08] ROBERTSON J., "Maximizing performance for higher k gate dielectrics", *Journal of Applied Physics*, vol. 104, p. 124111, 2008.

[SCA 07] SCAREL G., DEBERNARDI A., TSOUTSOU D., SPIGA S., CAPELLI S.C., LAMAGNA L., VOLKOS S.N., ALIA M., FANCIULLI M., "Vibrational and electrical properties of hexagonal La_2O_3 films", *Applied Physics Letters*, vol. 91, p. 102901, 2007.

[SCH 12] SCHAFFER M., SCHAFFER B., RAMASSE Q., "Sample preparation for atomic-resolution STEM at low voltages by FIB", *Ultramicroscopy*, vol. 114, pp. 62–71, 2012.

[SCH 09] SCHAMM S., COULON P.E., MIAO S., VOLKOS S.N., LU L.H., LAMAGNA L., WIEMER C.,TSOUTSOU D., SCAREL G., FANCIULLI M., "Chemical/Structural Nanocharacterization and Electrical Properties of ALD-Grown La_2O_3/Si Interfaces for Advanced Gate Stacks", *Journal of Electrochemical Society*, vol. 156, pp. H1–H6, 2009.

[SCH 11] SCHAMM-CHARDON S., COULON P.E., LAMAGNA L., WIEMER C., BALDOVINO S., FANCIULLI M., "Combining HRTEM–EELS nano-analysis with capacitance–voltage measurements to evaluate high-k thin films deposited on Si and Ge as candidate for future gate dielectrics", *Microelectric Engineering*, vol. 88, pp. 419–422, 2011.

[SCO 01] SCOTT A.J., BROWN A.P., NICHELLS A., BRYDSON R., "The effects of stray gun scattering in a commercial FEGTEM and its practical removal", *Inst. Phys. Conf. Ser.*, vol. 168, pp. 171–174, 2001.

[STE 01] STEMMER S., MARIA J.P., KINGON A.I., "Structure stability of La_2O_3/SiO_2 layers on Si(001)", *Applied Physics Letters*, Vol. 79, pp. 102–104, 2001.

[SUE 10a] SUENAGA K., KOSHINO M., "Atom-by-atom spectroscopy at graphene edge", *Nature*, vol. 468, pp. 1088–1090, 2010.

[SUE 10b] SUENAGA K., SATO Y., LIU Z., KATAURA H., OKAZAKI T., KIMOTO K., SAWADA H., SASAKI T., OMOTO K., TOMITA T., KANEYAMA T., KONDO Y., *Nature Chemistry*, vol. 1, pp. 415–418, 2010.

[TIM 98] TIMP G., BOURDELLE K.K., BOWER J., BAUMANN F., BOONE T., CIRELLI R., "Progress Toward 10nm CMOS", in *IEDM Technical Digest*, IEDM, San Francisco, pp. 615–618, 1998.

[TRI 04] TRIYOSO D.H., HEGDE R.I., GRANT J., FEJES P., LIU R., ROAN D., RAMON M., WERHO D., RAI R., LA L.B., BAKER J., GARZA C., GUENTHER T., WHITE B.E., Jr., TOBIN P.J., *Journal of Vacuum Science and Technology B*, vol. 22, pp. 2121–2127, 2004.

[TRI 05] TRIYOSO D.H., HEGDE R.I., GRANT J., SCHAEFFER J.K., ROAN D., WHITE B.E. JR., TOBIN P.J., *Journal of Vacuum Science and Technology. B*, vol. 23, pp. 288–297, 2005.

[TRI 08] TRIYOSO D.H., GILMER D.C., JIANG J., DROOPAD R., "Characteristics of thin lanthanum lutetium oxide high-k dielectrics", *Microelectronics Engineering*, vol. 85, pp. 1732–1735, 2008.

[TSO 08] TSOUTSOU D., SCAREL G., DEBERNARDI A., CAPELLI S.C., VOLKOS S.N., LAMAGNA L., SCHAMM S., COULON P.E., FANCIULLI M., "Infrared spectroscopy and X-ray diffraction studies on the crystallographic evolution of La2O3 films upon annealing", *Microelectronics Engineering*, vol. 85, pp. 2411–2413, 2008.

[TSO 09] TSOUTSOU D., LAMAGNA L., VOLKOS S.N., MOLLE A., BALDOVINO S., SCHAMM S., COULON P.E., FANCIULLI M., "Atomic layer deposition of $La_xZr_{1-x}O_{2-\delta}$ (x = 0.25) high-k dielectrics for advanced gate stacks ", *Applied Physics Letters*, vol. 94, p. 053504, 2009.

[WAL 03] WALTHER T., "Electron energy-loss spectroscopic profiling of thin film structures: 0.39nm line resolution and 0.04eV precision measurement of near-edge structure shifts at interfaces", *Ultramicroscopy*, vol. 96, pp. 401–411, 2003.

[WIL 01] WILK G.D., WALLACE R.M., ANTHONY J.M., "High-k gate dielectrics: Current status and materials properties Considerations", *Journal of Applied Physics*, vol. 89, no. 10, pp. 5243–5275, 2001.

[WON 06] WONG H., IWAI H., "On the scaling issues and high-j replacement of ultrathin gate dielectrics for nanoscale MOS transistors", *Microelectronics Engineering*, vol. 83, pp. 1867–1904, 2006.

[WU 02] WU X., LANDHEER D., QUANCE T., GRAHAM M.J., BOTTON G.A., "Characterization of $Hf_{1-x}Zr_xO_2$ gate dielectrics with 0 < x < 1 prepared by atomic layer deposition for metal oxide semiconductor field effect transistor applications", *Applied Surface Science*, vol. 200, pp. 15–20, 2002

Chapter 7

Characterization of Process-Induced Defects

We cannot think of a book dealing with transmission electron microscopy TEM and microelectronics that did not address the difficult field of "defects". Defects in crystalline silicon are often detrimental for devices notably affecting dopant diffusion during processing and finally causing leakage. A "good process" is often a process that creates fewer defects than others. In general, defects are created as a "reaction" of the crystals to strong "out-of-equilibrium" conditions generated by the processes used to fabricate the devices. More precisely, they provide a solution to the system to lower its energy, to relax, from a (too) high-energy state toward a lower energy configuration. Thus, it is important to know the crystallographic characteristics of the defects to understand their origin. Knowledge of their characteristics is also required to determine the image contrast rules to which they obey. This is a prerequisite condition to compare, on a quantitative basis, the effect of process options (e.g. annealing conditions) onto their sizes and densities, guiding the search for optimized process conditions. While classical techniques such as "weak-beam dark field" (WBDF) and "Fresnel contrast" imaging have been at the core of defect characterization for decades, new techniques making use of geometric phase analysis (GPA) of high resolution transmission electron microscopy (HRTEM) images or electron holography can provide complementary information, in particular the amplitude of Burgers vectors, as is demonstrated in the following. This chapter does not intend to replace existing books explaining in detail the theory of contrast applied to defects [WIL 96]. Its ambition is to describe, through examples, the techniques and methodologies that are to be used to study quantitatively *populations* of defects.

Chapter written by Nikolay CHERKASHIN and Alain CLAVERIE.

Making it probably too simple, we separate the defects into two classes: those that allow for stress relaxation between materials of different lattice constants or orientations, and those that result from the precipitation of supersaturated phases consisting of point defects and/or impurities. Both cases are addressed in the following section through selected examples often encountered when processing the electronic materials.

7.1. Interfacial dislocations

In general, imaging of dislocations is done under WBDF conditions. A general presentation of the technique can be found in [WIL 96]. Here, we recall the basics of the technique and provide examples of how it can be used in practical cases. Under dynamical bright-field (BF) or dark-field (DF) conditions, the image contrast originates from the strain fields surrounding the dislocations, which can extend over micrometers and thus overlap, rendering the detection and identification of individual dislocations often impossible. WBDF imaging of the same region consists of deviating the perfect crystal from exact Bragg conditions so that it is the bending of the atomic planes close to the core of the defects, which locally restores the Bragg conditions. In such images, the perfect crystal appears in gray/black whereas the regions close to the core of the defects diffract electrons and thus appear in white. Intuitively, we understand that as we increase this deviation from Bragg conditions s, the contrast characterizing the defect is located closer and closer to the core of the defect. This is to say that the contrast observed on a dislocation directly depends on the $\mathbf{g}$ vector used for imaging and on the sign and amplitude of s, which are determined from the associated diffraction pattern.

For a given set of imaging conditions, that is a diffraction vector $\mathbf{g}$ and a deviation from exact Bragg conditions, s, the intensity of the image of a dislocation line results from the differently weighted contributions of $|\mathbf{g} \cdot \mathbf{b}|$ and $|\mathbf{g} \cdot \mathbf{b} \otimes \mathbf{u}|$, where $\mathbf{u}$ is the unit vector tangential to the dislocation line [WIL 96]. The "true" extinction of contrast in the image is achieved when $|\mathbf{g}_1 \cdot \mathbf{b}| = 0$ and $|\mathbf{g}_1 \cdot \mathbf{b} \otimes \mathbf{u}| = 0$. If $|\mathbf{g}_2 \cdot \mathbf{b}| = 0$ but $|\mathbf{g}_2 \cdot \mathbf{b} \otimes \mathbf{u}| \neq 0$, the dislocation line will be seen in residual contrast. This "true" residual contrast does not depend on the sign of s, which allows us to distinguish it from the weak contrast observed for $|\mathbf{g}_2 \cdot \mathbf{b}| \ll 1$. Finally, the Burgers vector of a dislocation has to be perpendicular to the plane containing the two diffraction vectors, $\mathbf{g}_1$ and $\mathbf{g}_2$, and this plane has to be found experimentally.

Unfortunately, this method does not allow us to determine the *amplitude* of a Burgers vector nor does it provide information on the atomic structure of the defect.

This is where new techniques such as GPA of HREM images and dark-field electron holography (DFEH) (see Chapter 4) can provide supplementary information.

7.1.1. *Si(100)/Si(100) direct wafer bonding (DWB)*

Direct Si wafer bonding (DSWB) of different wafers has become an established technique in modern semiconductor technology, finding applications ranging from smart power integrated circuits to sensors [PAT 97, KIT 09]. DSWB consists of putting in contact the clean surfaces of two wafers and annealing until reconstruction of the interface occurs. In practice, the two wafers are never perfectly aligned so that there always exist a tilt angle and a twist angle between the two crystalline networks. As a result, DSWB interfaces reconstruct during annealing and produce networks of dislocations with characteristics depending on the surfaces in contact and on the twist and tilt disorientations between the two wafers [KIT 09, CHE 12]. Obviously, these dislocations alter the electrical characteristics of the bonded interfaces and, for this reason, it is not only essential to know their individual characteristics, but also their densities. A first difficulty arises from the need to image these interfaces, located at 300–500 mm from the surface of the DSWB structure, in plan-view, that is with the electron beam perpendicular to the interface plane. Indeed, while easy to prepare, cross-sectional images of the interface do not provide much information of the dislocation networks. There is no simple way to fabricate such plan-view samples. In general, very small angle beveling of the structure by tripod polishing, possibly followed by chemical etching, is a solution. However, as often, success will depend on experience and care, and going back and forth a numberal of times between the sample preparation room and the TEM will be necessary.

To illustrate the WBDF method, we show how the structure of the interface accommodating the twist angle between two bonded Si(001) wafers has been identified by Benamara *et al.* [BEN 95a, BEN 95b] following the classic procedure detailed in [WIL 96]. Such a type of interface is characteristic of a low angle grain boundary. In Figure 7.1, the same region of the interface is imaged using different diffraction vectors. In Figure 7.1(a), taken in BF along the [001] zone axis, all dislocation lines are visible. We note the presence of a square grid of dislocations and of broken lines disrupting the network (actually shifting the square grid by half of its periodicity). All other images are WBDF images taken under specific diffraction conditions. s is quantified through the conventional notation (g, ng), which indicates that the **g** vector is used for imaging while the Bragg condition is fulfilled for ng. This set of images shows that the vertical lines are invisible in the images taken with $\mathbf{g} = \bar{2}20(g,3g)$, (Figure 7.1(g)), with $\mathbf{g} = \bar{3}31(g,3g)$, (Figure 7.1(h)) and with $\mathbf{g} = \bar{1}11(g,5g)$ (Figure 7.1(i)). These **g** vectors thus satisfy the two relations: $|\mathbf{g}\cdot\mathbf{b}| = 0$ and $|\mathbf{g}\cdot\mathbf{b}\otimes\mathbf{u}| = 0$. Thus, the Burgers vector associated with these vertical dislocation lines lies along the [110] direction and the dislocations thus have a screw character. The horizontal lines are invisible in the images taken with $\mathbf{g} = 220(g,3g)$, (Figure 7.1(d)), with $\mathbf{g} = 331(g,3g)$, (Figure 7.1(e)) and with $\mathbf{g} = 11\bar{1}(g,5g)$ (Figure 7.1(f)) and thus their Burgers vector lies along the $[\bar{1}10]$ direction. These are screw dislocations as well.

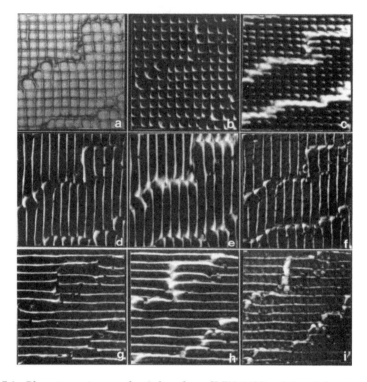

Figure 7.1. *Plan-view micrographs (taken from [BEN 95b]) of the dislocation network observed in Si(100)/Si(100) DWB interface taken under different diffraction conditions and allowing the identification of $\mathbf{b}_1$ (g, h and i) and $\mathbf{b}_2$ (d, e and f): (a) Bright-Field in [001] zone axis; (b)–(i) WBDF images of the same region taken with: (b) $\mathbf{g} = 400(g,2g)$; (c) $\mathbf{g} = 242(g,2g)$; (d) $\mathbf{g} = 220(g,3g)$; (e) $\mathbf{g} = 331(g,3g)$; (f) $\mathbf{g} = 11\bar{1}(g,5g)$; => $\mathbf{b}_2 = a/2[\bar{1}10]$; (g) $\mathbf{g} = 220(g,3g)$; (h) $\mathbf{g} = 331(g,3g)$; (i) $\mathbf{g} = 111(g,5g)$; => $\mathbf{b}_1 = a/2[110]$*

From this **g·b** analysis, it was demonstrated that the interface between two bonded silicon crystals with flat and parallel (001) surfaces showing a twist angle reconstructs via a square grid of pure screw dislocations with a Burgers vector, **b** of <110> type lying in the interface plane as schematized in Figure 7.2.

Characterization of Process-Induced Defects 169

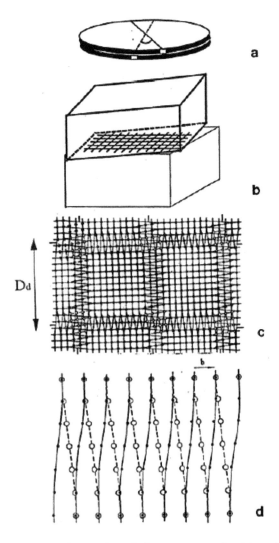

Figure 7.2. *Scheme showing how a tilt angle between two wafers (a) can be accommodated by a square grid (b) of screw dislocations of periodicity D_d (c). (d) It shows how the screw dislocations accommodate the disorientation*

A twist angle should be accommodated by the formation of screw dislocations with a Burgers vector equal to the interface translation vector along the dislocation lines. Its length, $a^{Si}b^{screw}$, can be expressed as a function of the twist angle, θ, and the average separation between dislocation lines, D, by:

$$a^{Si}b^{screw} = D \cdot 2\sin\frac{\theta}{2} \qquad [7.1]$$

Following this relation, Benamara *et al.* [BEN 95a, BEN 95b] have found that the amplitude of the Burgers vector of the pure screw <110> type dislocations accommodating the twist angle between the two (001) wafers was equal to 1/2.

Following a similar procedure, it has been demonstrated that the tilt angle at the interface resulting from the two miscut angles of the wafers is accommodated by generating 60° dislocations, that is equally spaced atomic steps at the interface [BEN 95a, BEN 95b]. More recently, the same procedure was used to determine the type and the amplitude of the Burgers vector of the dislocations formed at the interface, and accommodating a twist angle between Si(110) and Si(100) wafers along their common [110] direction [CHE 12]. Such an interface is characteristic of a high angle grain boundary. This interface is today of particular interest because these DWB structures are being considered as starting materials to fabricate pFET and nFET using the "hybrid orientation technology" approach. Dislocations formed to accommodate a twist angle between Si(110) and Si(100) wafers along their common [110] direction have also been found to be of the pure screw type. However, their amplitude was found to be twice as small as that of the dislocations formed at the Si(001)/Si(001) interface, resulting in a twice as small distance between dislocations [CHE 12].

7.1.2. *SiGe heterostructures*

Mismatched epitaxial layers are widely used in modern semiconductor technology. For example, compressively strained germanium can be used as channel material with increased hole mobility in state-of-the-art p-type metal oxide semiconductor field effect transistors. Strained Ge layers can be grown epitaxially on fully relaxed, high Ge content SiGe virtual substrates [HAR 10]. However, misfit dislocations will form to relax the elastic energy stored in the strained overlayer if its thickness exceeds a critical value. We show here the example of a 67-nm-thick Ge layer epitaxially grown onto a $Si_{0.15}Ge_{0.85}$ virtual substrate. As a result of misfit accommodation, a rectangular network of misfit dislocations is generated close to the Ge/SiGe interface, with a rather low linear density of 5.7×10^4 cm^{-1} and an average distance between dislocations, <d> $\cong$ 350 nm [HAR 10]. However, using the "classical" method described above, the characteristics of these dislocations could not be determined. Indeed, whatever the diffraction vector used for imaging, the extinction of the contrast associated with these dislocation lines cannot be observed (Figure 7.3).

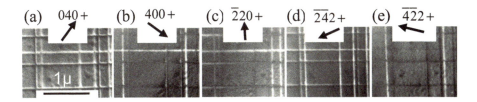

Figure 7.3. *WBDF plan view (001) images of the misfit dislocation network found close to the Ge/SiGe interface. Note that all the dislocation lines are visible whatever the diffraction vector*

At this stage, it can be suspected that this results from the presence of pairs of dislocations located close to each other and having different Burgers vectors. Such pairs of dislocations are often observed in epitaxial layers and result from the dissociation of the misfit dislocations into partial dislocations. When the dissociation length is very small (of the order of 1–2 nm), their contrasts superimpose, rendering defect analysis impossible. The recently invented DFEH technique, presented in Chapter 4, can be used to solve this problem.

Figures 7.4(a) and (b) show the phase images extracted from two DFEH images taken with $\mathbf{g}_1 = \bar{1}\bar{1}1$ and $\mathbf{g}_2 = 111$ from a cross-sectional specimen of the structure. From these images, the in-plane strain, ε_{xx}^{SiGe} (Figure 7.4(c)) and out-of-plane strain ε_{zz}^{SiGe} (Figure 7.4(d)) relative to $Si_{0.15}Ge_{0.85}$ have been extracted.

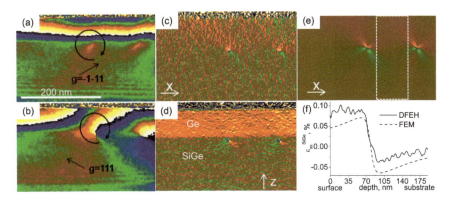

Figure 7.4. *Phase images extracted from DFEH images of the $Ge/Si_{0.15}Ge_{0.85}$ structure observed on a (110) cross-sectional sample, for $\mathbf{g}_1 = \bar{1}\bar{1}1$ (a) and for $\mathbf{g}_2 = 111$ (b). Deduced in-plane strain, ε_{xx}^{SiGe}, (c) and out-of-plane strain, ε_{zz}^{SiGe} (d). FEM simulation of in-plane strain, ε_{xx}^{SiGe} (e). Line profiles along depth obtained from (c) and (e) when averaging in the rectangular region located between two dislocations (marked in white in (e))*

Having a lattice parameter larger than that of the SiGe alloy, the Ge layer appears in red. Two dislocations located close to the Ge/SiGe interface are seen edge-on in the right-hand part of the strain images (c) and (d). The projection in the $(1\bar{1}0)$ image plane, $\mathbf{b}_p$, of the Burgers vector, $\mathbf{b}$, of the $\begin{bmatrix}1\bar{1}0\end{bmatrix}$ dislocation line can be determined following the procedure described in [HŸT 03] and applied to an edge dislocation. By definition, the Burgers vector of a dislocation is the integral of the strain surrounding its core. Thus, we write:

$$\mathbf{b}_p = -\frac{1}{2\pi}(\Delta P_{g1} \cdot \mathbf{a}_1 + \Delta P_{g2} \cdot \mathbf{a}_2) \qquad [7.2]$$

where ΔP_{g1} (ΔP_{g2}) is the phase shift obtained around the dislocation core for $\mathbf{g}_1$($\mathbf{g}_2$). $\mathbf{a}_1$ ($\mathbf{a}_2$) is the real space vector corresponding to $\mathbf{g}_1$ ($\mathbf{g}_2$).

Since $\mathbf{g}_1 = \bar{1}\bar{1}1, \mathbf{a}_1 = 1/4[\bar{1}\bar{1}2], \Delta P_{g1} = 0$ and $\mathbf{g}_2 = 111, \mathbf{a}_2 = 1/4[112], \Delta P_{g2} = -2\pi$, we obtain $\mathbf{b}_p = 1/4[112]$.

Note that the two dislocations seen at a distance of 125 nm from each other in Figures 7.4(c) and (d) have the same Burgers vector *projection*. Both dislocations might be perfect 60° dislocations with full Burgers vector $\mathbf{b} = 1/2[101]$. Another projection obtained either by tilting the sample or preparing a new sample cut along a different direction would be required to prove it. However, if a dislocation line is no longer seen edge-on in the image, the interpretation of the phase images will be much more complicated. This is clearly a disadvantage of this method. While WBDF imaging unambiguously provides the Burgers vector direction (but not the amplitude) of even complex, mixed-type dislocations, DFEH allows us to determine the amplitude of only the projection of this Burgers vector in the image plane. However, the complementarity of both techniques is evident in this example.

7.2. Ion implantation induced defects

Ion implantation is a classical technology in the modern semiconductor industry. It is used, for example, for doping active regions of transistors and for fabricating SOI wafers [BRU 99]. The ions slow down by interacting with the electron clouds at high energy and then by displacing atoms from the crystalline target once their energy is reduced. For this reason, the "crystalline damage" is buried between the surface and a certain depth. This primary, collisional damage consists of impurity atoms, self-interstitials and vacancies that evolve, while recombining with one another, toward structures stable at the wafer temperature. At room temperature, only I_2, V_2 and impurity complexes involving one or the other type of point defect exist. When increasing the ion dose, an amorphous layer can form at all depths where the concentration of defects is higher than a certain threshold [KOF 09a]. However, defects survive above and below this amorphous layer. When increasing the temperature, these clusters and complexes dissociate and recombine to finally condense into larger defects visible by TEM. During annealing, these defects grow by interchanging the Si atoms or vacancies they contain through a competitive mechanism named as Ostwald ripening [CLA 03]. These fluxes of Si atoms and vacancies are at the origin of diffusion anomalies during processing.

In the following section, we will show how "old" and "new" techniques can be used to identify these defects. Then, knowing their crystallographical characteristics, the image conditions under which their populations can be statistically analyzed can be determined. We separate defects into two classes: those formed by the agglomeration of "excess atoms", that is Si self-interstitials and/or impurities (traditionally named extrinsic) and those formed by the condensation of vacancies (traditionally named intrinsic).

7.2.1. *Defects of interstitial type*

7.2.1.1. *Strain and defects found after implantation (not annealed)*

As ion implantation consists of adding atoms within a solid, one of the important characteristics of every implantation is the appearance of an out-of-plane strain in the implanted region with a depth-distribution and amplitude strictly depending on implantation conditions. Actually, this microscopic strain results from the overlapping/averaging of nanoscopic strain fields generated by individual defects in the substrate [MIC 03, SOU 06]. Measuring the deformation depth-distribution helps in quantifying the strain energy of the system and in understanding the role and type of the observed defects. In the following, we present two examples of strain depth-distributions determination using the DFEH technique, one obtained after an amorphizing implant and the other after a non-amorphizing implant.

A typical example of an amorphizing implant is provided by a Ge wafer implanted with Ge^+ ions at 150 keV and with a fluence of 5×10^{14} ions/cm^2. These implantation conditions have been studied in detail, to end a long-standting controversy on the possible existence of "EOR" defects in Ge [KOF 09b]. Figure 7.5(a) shows an XTEM image taken in WBDF conditions of this sample. Under these diffraction conditions, the deep crystal appears in dark (not diffracting), the amorphous layer in gray (diffusely scattering) while the c/a interface appears in white in the image. This is the first experimental evidence that the crystal is heavily strained immediately below this interface.

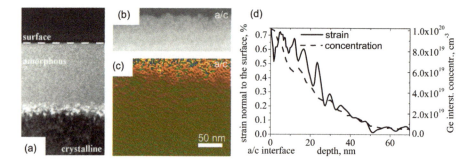

Figure 7.5. *(a), WBDF-XTEM image of an amorphous Ge layer formed by a 150 keV Ge^+ 5 $x10^{14}$ ions/cm^2 implantation (taken from [SOU 06]); (b) Same region but imaged using exact DF-XTEM conditions; (c) strain mapping ε_{zz} of the lattice normal to the wafer surface obtained by DFEH; (d) left axis, strain ε_{zz} and right axis, concentration of "excess Ge interstitials" (calculated by Monte Carlo method) as functions of the distance from the c/a interface)*

A DF image of the very same region but taken classically under exact Bragg conditions does not show such a contrast but can be used to measure the roughness of this interface (Figure 7.5(b)). To quantify this strain, we make use of the DFEH technique. Being diffraction based, it provides information on the strain occurring only on the crystalline side of the interface. Figure 7.5(c) shows the mapping of the strain perpendicular to the c/a interface. This image is obtained from the hologram taken using $g = 004$. The mean variation of this strain field, ε_{zz}, as a function of the distance from the c/a interface can be extracted by numerical integration of this image along the interface plane. Figure 7.5(d) shows the result. It is clear that the lattice is positively strained in this direction and that the amplitude of this strain quickly decreases from a maximum of 0.6–0.7% close to the c/a interface to below the detection limit (<0.05%) at a depth of 50 nm from this interface. Interestingly, these characteristics have also been reported by Bisognin *et al.* [BIS 08] using high-resolution X-ray diffraction (XRD) on the same samples. No deformation was detected either by DFEH or by XRD in the direction parallel to the wafer surface. Actually, this strain profile mimics the concentration profile of the Ge interstitial atoms recoiled below the c/a interface by the implantation and it has been suggested that these interstitial atoms, isolated or in the form of small clusters, generate this strain [KOF 09b, BIS 08]. During annealing at 300–400°C, extended (interstitial) defects are detected using WBDF techniques [KOF 09b], which confirms the hypothesis. We note with this example that the DFEH technique has allowed us to detect defects too small to be imaged individually through the imaging of the strain field that their *population* generates.

Hydrogen ion implantation in silicon provides a good example of the interplay between stress and defects in the case of a non-amorphizing implant. Hydrogen implantation is used to slice and finally transfer thin Si layers from a donor substrate onto a host material, providing a versatile method for the fabrication of SOI wafers and heterostructures [BRU 99, ASO 01]. The separation of the top layer from the donor substrate is achieved during annealing by a controlled fracture; an outcome of a complex microstructure evolution in which stress and strain play crucial roles [REB 09, HEB 07, PER 08]. Indeed, the implantation of H ions creates a buried damage layer in the target substrate consisting of gas atoms, self-interstitials and vacancies, arranged in different forms of hydrogen-vacancy and hydrogen-interstitial complexes [PER 08]. During annealing, these complexes recombine and finally "platelets", two-dimensional precipitates formed by the agglomeration of vacancies and H atoms and molecules, nucleate. These defects are studied in detail later on in this chapter. We focus here on the characterization of strain in these H-implanted Si wafers before annealing, a prerequisite step to understand the driving force acting during annealing.

Strains in general and out-of-plane strain in particular can be measured by XRD [SOU 06]. This technique provides accurate values of the strain at different depths, but no absolute information on their depth-distribution. Moreover, it has no spatial resolution required to understand the origin of this strain by correlating it with possibly nanoscopic defects. These limitations are totally relaxed when using DFEH.

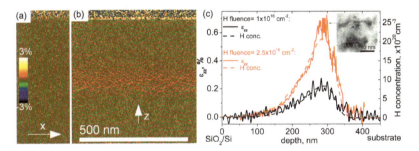

Figure 7.6. *Si(001) wafer implanted with 1×10^{16} H at./cm². Cross-sectional sample observed along <110>. (a) In-plane, ε_{xx}, and (b) out-of-plane, ε_{zz}, strain maps obtained by DFEH; (c) vertical strain profiles of the out-of-plane ε_{zz} strain component obtained by integration of the strain maps. Samples implanted with a fluence of 1×10^{16} H at./cm² (black solid line) and with 2.5×10^{16} H at./cm² (red solid line). The corresponding H depth distributions obtained by SIMS are also shown (dashed lines). The BF image in the insert shows that platelets are detected at depths where the H concentration is maximum only after 2.5×10^{16} H at./cm² implantation*

In Figure 7.6, we show the strain mappings obtained by DFEH on a cross-sectional (110) Si(001) sample implanted with 1×10^{16} H⁺/cm² at 37 keV. The wafer was covered with 145-nm-thick SiO_2 layer before implantation. Figures 7.6(a) and (b) have been extracted from $\mathbf{g}_1 = 1\bar{1}1$ and $\mathbf{g}_2 = \bar{1}11$ DF holograms, respectively.

Figure 7.6(a) shows that the crystal is not deformed in the direction parallel to the surface. Alternatively, the out-of-plane strain is positive (Figure 7.6(b)) and its integrated profile (Figure 7.6(c), black solid line) perfectly follows the concentration profile of hydrogen experimentally measured by SIMS (Figure 7.6(c), black dashed line). This strain is thus the result of a complex reaction of the matrix crystal submitted to stresses generated by individual defects such as H atoms alone or associated with self-interstitials or vacancies within complexes [PER 08]. The "macroscopical evidence" is that the out-of-plane strain is proportional to the hydrogen content in the layer, a classical result for "dilute interstitials" systems.

However, the situation may become different when increasing the H concentration. Figure 7.6(c) also shows the out-of-plane strain profile (red solid line) obtained on a sample implanted to a higher fluence of 2.5×10^{16} H^+/cm^{-2}, and which is again compared to the SIMS profile. At low concentration, we find again the same proportionality but, close to the peak, the amplitude of the strain is larger than expected, largely overcoming the SIMS profile. Conventional TEM imaging of this region reveals the presence of extended defects (platelets, see insert in Figure 7.6(c)). Although we do not discuss this effect at length in this chapter, this demonstrates that the presence or absence of extended defects drastically affects the amplitude of the strain that is generated by ion implantation.

The combination of conventional and holographic techniques of the same regions of specimens provides a unique ensemble of information necessary to address and understand the difficult interplay between defects and stress/strain found in silicon and most materials and structures during their processing.

7.2.1.2. Defects found after implantation and annealing

In general, the extended defects observed after annealing of ion-implanted silicon are of (self) interstitial type, i.e. composed of Si interstitial atoms. These interstitials can either be generated by the activation of the doping impurity, that is by its passage onto a lattice site through the ejection of a Si atom, or accumulated after recoiling below the amorphous/crystalline interface in the case of amorphizing implants [CLA 95, LAA 95]. During annealing, these self-interstitial clusters grow in size by interchanging the Si atoms they are composed of [CLA 03]. A particular characteristic of interstitial Si precipitates in Si is that they change morphology as they grow in size. Initially stored as di-interstitials (I_2) stable at room temperature, during annealing, the Si atoms condense as small clusters consisting of several atoms [COW 99] that transform into {113} rod-like (RL) defects [DAG 95], the latter transforming into {111} RL defects [BON 06], themselves finally transforming into {111} dislocations loops (DLs) of two types [DE 94, BON 98]. Except for small clusters, the structure of all other defects can be approximated by extrinsic-type DLs of different shapes, habit planes and Burgers vector.

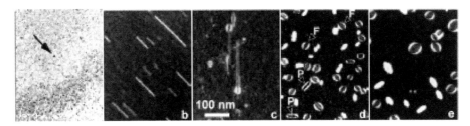

Figure 7.7. *The different types of interstitial defect that can be observed in ion-implanted Si depending on annealing conditions: small clusters (a), {113} RL defects (b), {111} RL defects transforming into DLs (c), mixed population of perfect and faulted dislocation loops (d) and faulted (Frank) loops. All images taken under WBDF conditions on plan-view samples*

The growth laws and energetics of this complex Ostwald ripening phenomenon have recently been determined [CLA 03] and the resulting master equations have been integrated into modern process simulators that are now able to predict defect evolution and dopant diffusion during annealing [ZOG 07]. It is interesting to recall the pathway taken by the main contributors of this success along their investigations. Dealing with various types of defect of different sizes, sometimes simultaneously present in the same sample, a first step is to formally identify these defects, that is to find their crystallographical characteristics, deduce the number of orientational variants and determine the image contrast rules they obey. A second step is to deduce the imaging conditions suitable for the quantitative and statistical analysis of defects populations, in terms of size-distributions, densities and number of Si atoms they contain.

In the following section, we focus on the most famous of these defects, namely the {113} RL defects [DAG 95]. We recall the different techniques and methodologies that have been used to identify them and to analyze their evolution during annealing.

7.2.1.3. {113} RL defects

The typical population of {113} RL defects we study here were formed by implanting Si(001) wafers with 2×10^{14} cm^{-2} Si$^+$ ions at 100 keV followed by annealing at 850°C during 30 s in flowing N_2.

7.2.1.3.1. Type and projection of a Burgers vector by GPA of HRTEM images

Figure 7.8(a) shows a cross-sectional HRTEM image of a {113} RL defect that is seen edge-on along the (110) direction. The width of this defect is about 4–5 nm. Inverse Fourier filtering of this image can be used to suppress the contributions from the Si matrix (see inset in Figure 7.8(a)). Such processing of this HREM image proves that the defect consists of chains of di-interstitials lying on the $(1\bar{1}3)$ plane and extending along the [110] direction, that is perpendicularly to the image plane. This atomic structure is similar to the structure found for {113} planar defects that are of much larger sizes, truly two-dimensional and only observed after electron irradiation in Si [TAK 91a].

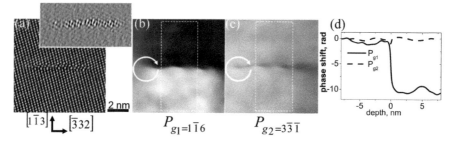

Figure 7.8. *Cross-sectional (110) HRTEM image of a $(1\bar{1}3)$ RL defect seen edge-on (a). Insert in (a) shows the inverse Fourier filtered imaged obtained after elimination of the Si matrix components. GPA of the HRTEM image: phase images obtained using $g_1 = 1\bar{1}6$ (b), and using $g_1 = 33\bar{1}$ (c). Line profiles extracted from (b) and (c) by integration along the defect habit plane and over the white dotted rectangles seen on the phase images (d)*

The GPA of this HRTEM image can be used to obtain the phase images corresponding to two perpendicular g vectors, $\mathbf{g}_1 = 1\bar{1}6$ (Figure 7.8(b)) and $\mathbf{g}_1 = 33\bar{1}$ (Figure 7.6(c)), following the procedure described in [HŸT 03]. Let us first remark that, whatever the diffraction vector used, there is no phase jump when circling around the defect. This indicates that the defect hosts two dislocations with opposite Burgers vectors, as DLs show. To characterize this defect, we focus on the phase profile taken perpendicularly to the defect habit plane (Figure 7.8(d)). The phase jump across the defect plane is zero for $\mathbf{g}_1 = 33\bar{1}$ and equals approximately -10 rad for $\mathbf{g}_1 = 1\bar{1}6$. The negative value of this phase jump for $\mathbf{g}_1 = 1\bar{1}6$ demonstrates the extrinsic nature (interstitial type) of the DL. Again, applying equation [7.1], the projection of the Burgers vector of the $(1\bar{1}3)$ RL defect in the (110) image plane can be calculated, yielding $\mathbf{b}_p = 1/24 \pm 1[1\bar{1}6]$. This value is very close to that obtained for the full Burgers vector of the well-known {113} planar defect [TAK 91a, TAK 91b]. It is thus very probable, although not demonstrated, that both defects have the same Burgers vector, lying in the (110) plane. If so, having the same atomic structure and Burgers vectors, the density of interstitial Si atoms, which both defects contain, should be about the same, that is about 5 nm^{-2}. Thus, the {113} RL defect imaged in Figure 7.8 can be seen as an extrinsic DL lying on the $(1\bar{1}3)$ habit plane, elongated along the [110] direction parallel to the (001) wafer surface.

7.2.1.3.2. Crystallography of {113} defects

Because of the cubic (3 mm) symmetry of the silicon crystal, there exists six possible <110> directions for the defect elongation and 12 possible {113} habit planes. Among them, only two contain one of the six equivalent directions. Thus, there are 12 orientational variants of these {113} RL defects, all elongated along one of the six possible <110> directions (see Figure 7.9(a)). Four variants such as $R^3_{1\text{and}4}$ and $R^6_{2\text{and}3}$ are elongated in a plane parallel to the (001) wafer surface and we name them "parallel" defects. The other eight variants such as $R^1_{1\text{and}2}, R^2_{1\text{and}3}, R^4_{3\text{and}4}, R^5_{2\text{and}4}$ are all elongated along directions inclined at 45° with respect to the wafer surface and thus we name them "inclined" defects.

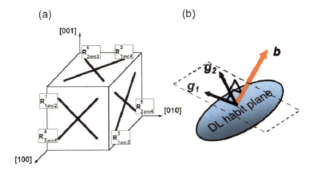

Figure 7.9. *(a) Visualization of the elongation axes **u** of the "parallel" (R^3 and R^6) and "inclined" (R^1, R^4 and R^5, R^2) variants of {113} RL defects in the three {100} planes. (b) Scheme showing the habit plane of a DL, its Burges vector and the plane containing the diffraction vectors perpendicular to its Burgers vector*

In the following section, we show how the "classical" method based on the contrast analysis of WBDF images, previously used to identify large and isotropic {111} DLs in ion-implanted Si [DE 94], can be implemented for the small elongated {113} RL defects. This method allows us to identify the nature of the defect (extrinsic/intrinsic), its habit plane and the direction (but not the amplitude) of its Burgers vector.

7.2.1.3.3. Burgers vector direction using WBDF images

The direction of the full Burgers vector of a DL can be determined using the same visibility criterion as used for a single dislocation line. There are three possible types of contrast arising from a DL: (1) a strong/weak contrast for ($|\mathbf{g}\cdot\mathbf{b}| \geq 0.4 / |\mathbf{g}\cdot\mathbf{b}| < 0.4$); (2) a residual contrast for ($|\mathbf{g}\cdot\mathbf{b}| = 0, |\mathbf{g}\cdot\mathbf{b}\otimes\mathbf{u}| \neq 0$); and (3) no or zero contrast for ($|\mathbf{g}\cdot\mathbf{b}| = 0, |\mathbf{g}\cdot\mathbf{b}\otimes\mathbf{u}| = 0$). We then must find two diffraction vectors, $\mathbf{g}_1$ and $\mathbf{g}_2$, giving rise to zero (true extinction) or residual contrasts of the DL in the image. Once found, the normal to the plane defined by these two vectors is parallel to the Burgers vector of the defect (Figure 7.9(b)). For a circular DL, both the zero contrast and the residual contrast can co-exist in different parts of the DL. The residual contrast does not depend on the sign of s. If the contrast of a DL is weak but depends on the sign of s, then the diffraction vector that has been chosen is not perpendicular to the Burgers vector ($|\mathbf{g}\cdot\mathbf{b}| \ll 1$). If the contrast of the DL is quite strong but does not depend on the sign of s, then the chosen diffraction vector is really perpendicular to the Burgers vector ($|\mathbf{g}\cdot\mathbf{b}| = 0$ but $|\mathbf{g}\cdot\mathbf{b}\otimes\mathbf{u}|$ is rather large). In both cases, the key to distinguish the strong/weak residual contrast from the strong/weak contrast is to concentrate on the behavior of this contrast when changing the sign of s.

U	B	Plane	name	$\mathbf{B} \cong [001]$			$\mathbf{B} \cong [\bar{1}\bar{1}3]$		$\mathbf{B} \cong [116]$
				$\mathbf{g} = 040$	$\mathbf{g} = [220]$	$\mathbf{g} = [\bar{2}20]$	$\mathbf{g} = [\bar{2}42]$	$\mathbf{g} = 422$	$\mathbf{g} = [\bar{3}31]$
[01$\bar{1}$]	[6$\bar{1}$1]	(3$\bar{1}$1)	R_1^1	w/i	s/i	s/o	s/o	s/o	s/o
	[$\bar{6}$11]	($\bar{3}$11)	R_2^1	w/i	s/o	s/i	w/o	s/o	s/i
[10$\bar{1}$]	[161]	($\bar{1}$31)	R_1^2	s/i	s/i	s/i	s/o	s/o	s/o
	[$\bar{1}$61]	(1$\bar{3}$1)	R_3^2	s/o	s/o	s/o	s/o	w/o	s/i
[1$\bar{1}$0]	[116]	(11$\bar{3}$)	R_1^3	w/i	w/i	0	s/i	s/i	r
	[$\bar{1}\bar{1}$6]	($\bar{1}\bar{1}$3)	R_4^3	w/o	w/o	0	w/i	w/i	s/i
[011]	[6$\bar{1}\bar{1}$]	(3$\bar{1}\bar{1}$)	R_3^4	w/o	s/i	s/o	s/i	s/i	s/o
	[$\bar{6}\bar{1}\bar{1}$]	($\bar{3}\bar{1}\bar{1}$)	R_4^4	w/o	s/o	s/i	s/o	s/o	s/i
[101]	[$\bar{1}$61]	($\bar{1}$31)	R_2^5	s/i	s/i	s/i	s/i	s/i	s/o
	[$\bar{1}\bar{6}\bar{1}$]	($\bar{1}\bar{3}\bar{1}$)	R_4^5	s/o	s/o	s/o	s/o	s/o	s/i
[110]	[$\bar{1}$16]	($\bar{1}$13)	R_2^6	w/i	0	w/i	s/i	s/i	w/i
	[$\bar{1}\bar{1}$6]	($\bar{1}\bar{1}$3)	R_3^6	w/o	0	w/o	s/i	s/i	w/i

Table 7.1. *The {113} RL defects distributed into 12 variants (left side). Contrast predictions are shown for s < 0 and different g. $\mathbf{B} \cong [hkl]$ is the beam direction close to the ZA $[hkl]$. s-strong contrast ($|\mathbf{g}\cdot\mathbf{b}| \geq 0.40$), w-weak contrast ($|\mathbf{g}\cdot\mathbf{b}| < 0.40$), r-residual contrast ($|\mathbf{g}\cdot\mathbf{b}| = 0, |\mathbf{g}\cdot\mathbf{b}\otimes\mathbf{u}| \neq 0$), 0-zero contrast ($|\mathbf{g}\cdot\mathbf{b}| = 0, |\mathbf{g}\cdot\mathbf{b}\otimes\mathbf{u}| = 0$), o and i-outside and inside contrast*

The prediction of contrast (strong, weak, residual and zero) for the 12 variants of the {113} RL defect, calculated for **g** = 040, 220, $\bar{2}20$, 242, 422 and $\bar{3}\bar{3}1$, and various beam directions are shown in Table 7.1.

Figure 7.10 shows a series of WBDF images of the very same region of a plan-view sample containing RL defects.

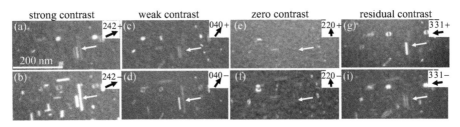

Figure 7.10. *Series of WBDF-PVTEM images taken with different* **g** *and s demonstrating that the RL defect marked by the white arrow has its Burgers vector along [116]. The sign "+/– indicates the positive/negative sign of s*

The defect pointed out by the white arrow shows strong contrast for $\mathbf{g} = 242, s < 0, \mathbf{B} \cong [\bar{1}\bar{1}3]$ ($|\mathbf{g}\cdot\mathbf{b}| = 0.72, |\mathbf{g}\cdot\mathbf{b}\otimes\mathbf{u}| = 0.91$) (Figure 7.10(b)), weak contrast for $\mathbf{g} = 040, s < 0, \mathbf{B} \cong [001]$ ($|\mathbf{g}\cdot\mathbf{b}| = 0.16, |\mathbf{g}\cdot\mathbf{b}\otimes\mathbf{u}| = 0.68$) (Figure 7.10(d)), zero contrast for 001 $\mathbf{g} = \bar{2}20, s < 0, s > 0, \mathbf{B} \cong [001]$ ($|\mathbf{g}\cdot\mathbf{b}| = 0, |\mathbf{g}\cdot\mathbf{b}\otimes\mathbf{u}| = 0$) (Figures 7.10(f) and (e)) and residual contrast for $\mathbf{g} = \bar{3}\bar{3}1, s < 0, s > 0, \mathbf{B} \cong [116]$ ($|\mathbf{g}\cdot\mathbf{b}| = 0, |\mathbf{g}\cdot\mathbf{b}\otimes\mathbf{u}| = 1.07$) (Figures 7.10(i) and (g)).

This behavior demonstrates that this RL defect, elongated along the $[\bar{1}10]$ direction, has its Burgers vector along the [116] direction.

7.2.1.3.4. Defect type: the "inside/outside" contrast

Finally, two questions remain: (1) is the {113} RL defect an extrinsic (interstitial-type) defect or an intrinsic (vacancy-type) defect? and (2) what is its habit plane? Although we can think that both answers can be guessed from the result of the GPA analysis of HRTEM image shown in a previous section, this is not formally true. Actually, the analysis required to identify the type and habit plane of a DL or RL defect is somehow fastidious. It makes use of a complete analysis of the contrast observed on WBDF images of the defect under specific conditions and the possibilities to make mistakes are numerous. We should then work with extreme care, with notations and with the microscope.

Figure 7.11(a) shows an analog of the famous scheme drawn in the 1960s by Groves and Kelly [GRO 61] and to be used for the identification of the nature of DLs. It provides the contrast that will be seen in the image of an interstitial DL depending on imaging conditions (electron beam direction -**B**, diffraction vector **g**, sign of the angular deviation from exact Bragg condition s), its habit plane and the direction of its Burgers vector [DE 94, GRO 61].

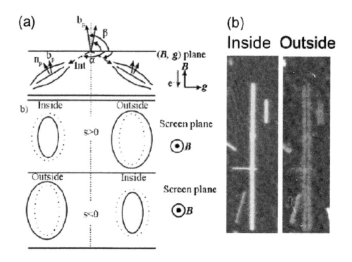

Figure 7.11. *(a) Schematic diagram in the (**B**, **g**) plane used to determine the inside/outside contrast of an interstitial dislocation loop in a WBDF image obtained with a given (**g**, **B**, s) set. **Int** is the downward-directed intersection of the defect habit plane and the (**B**, **g**) plane; **n**$_p$ is the upward normal to **Int**; **b**$_p$ is the projection of the Burgers vector, **b**, on the (**B**, **g**) plane. **b** should be indexed following the Finish–Start/Right-Handed (FS/RH) definition of **b** with a clockwise positive direction around the defect as seen on the screen [33]; α is the clockwise angle between **g** and **Int**; β is the anticlockwise angle between **g** and **b**$_p$. The "as seen on the screen" image (solid line circles) deviates from the projection of the core of the DL (dashed line circles) depending on the sign of s and on α ($\alpha > \pi/2$ (left column) or $\alpha < \pi/2$ (right column)). (b) WBDF images of the same {113} RL defects seen in inside (left) or outside (right) contrast*

The physical origin of the inside–outside contrast effect lies in the inversion of the sign of s at the top and bottom edges of the loop due to the strong bending of the crystalline matrix surrounding it. It causes the displacement of the image of the DL (solid line) with respect to the projection of the core of the DL (dashed line), which increases as s increases, and extends toward the side giving maximum intensity of diffracted electrons (see, e.g. [WIL 96] for more details). An example of inside–outside contrasts is shown in Figure 7.11(b) for a {113} defect observed under WBDF conditions providing the same intensity contrast ($|\mathbf{g} \cdot \mathbf{b}| \geq 0.40$) and for two

$s \neq 0$ of opposite signs. In these images, the contrast of the DL lies either inside or outside the "real" position of the DL core.

To provide the correct indexation of all vectors, the following conventions are to be used:

1) right-handed coordinate system is assumed for the crystal;

2) **B** is opposite to the electron beam in the microscope;

3) **g** is directed from the left to the right;

4) s has a positive value when the center of the reciprocal lattice streak lies inside Ewald's sphere;

5) **Int** is the downward-directed intersection of the defect habit plane and the (**B**, **g**) plane;

6) $\mathbf{n_p}$ is the upward normal to **Int**;

7) $\mathbf{b_p}$ is the projection of the Burgers vector, **b**, on the (**B**, **g**) plane. The scheme refers to the Finish–Start/Right-Handed (FS/RH) definition of **b** with a clockwise positive direction around the defect as seen on the screen [EDI 75];

8) α is the clockwise angle between **g** and **Int**, and β is the anticlockwise angle between **g** and $\mathbf{b_p}$;

9) a positive tilt of the sample around any vector is defined by the usual rule of a right-handed screw;

10) the indexation of **g** vectors is carried out from the lower hemisphere stereographic projection. The lower hemisphere of the projection is used in order to keep the same sense of the displacement of Kikuchi lines and poles on the stereogram and on the screen of the microscope when tilting the specimen. For example, by tilting the sample by $+25°$ around the $[\bar{1}10]$ direction, the Kikuchi lines move from the right to the left of the screen, while the $[1\bar{1}3]$ pole in the map is put in the center, that is parallel to the electron beam. Thus, the upward **B** should be indexed as $[\bar{1}\bar{1}3]$.

The selection of the beam direction **B** to be used for adequate WBDF imaging conditions can be done as follows. For a given **g**, we have to slightly deviate the sample from a high-symmetry zone axis by combining two separate tilts. The first tilt (~4°), around the **g** axis, should approximately maintain two-beam conditions. The second tilt, athwart to **g**, is used to tune the sense and the magnitude of s. If only the first tilt angle is used, then two opposite reflections, +**g** and –**g**, will be equally excited and thus s will be negative. Exact indexes of the actual **B** can be calculated and the sign of s can be determined from the position of the g-Kikuchi white line observed in the diffraction pattern.

The scheme was initially intended to predict the inside–outside contrast generated by "pure edge" DLs, that is DLs having their Burgers vector perpendicular to their habit plane for which $\alpha,\beta < \pi/2$ or $\alpha,\beta > \pi/2$. In such a case, the contrast only depends on the sign of the **g·b**$_p s$ product. Actually, similar contrast is expected for $\alpha,\beta < \pi/2$ and $\alpha < \pi/2, \pi/2 < \beta < \pi - \alpha$ or for $\alpha,\beta > \pi/2$ and $\alpha > \pi/2, \pi - \alpha < \beta < \pi/2$.

Thus, the method and the scheme shown in Figure 7.11 are also valid for mixed-type DLs. Finally, after the proper indexation of all vectors in Figure 7.11(a), for $s < 0$, the contrast of an interstitial-type DL will depend on the values of the α and β angles: inside contrast for $\{\alpha < \pi/2, -\alpha < \beta < \pi - \alpha\}$, outside contrast for $\{\alpha > \pi/2, 2\pi - \alpha < \beta < \pi - \alpha\}$. Obviously, the contrast would reverse for vacancy-type DLs.

By comparing the contrast of a DL observed under given imaging conditions with the predictions, the nature of a DL can be determined provided that both the habit plane and the direction of its Burgers vector are known. The comparison between contrast predictions shown in Table 7.1 and the images shown in Figure 7.11(b) taken with **B** close to the $[\bar{1}1\bar{3}]$ direction, **g** = 242 and $s < 0$, demonstrates that the {113} RL defects are extrinsic, that is of interstitial type.

When the habit plane of a defect is uncertain, for fixed β and inside (or outside) contrast, different solutions are possible as far as α lies in the range $\alpha > \pi/2$ (or $\alpha < \pi/2$). This ambiguity can be overcome noting that the DL contrast should reverse as soon as α passes from an angle slightly smaller than $\pi/2$ (or π) to an angle slightly higher than $\pi/2$ (or 0) (Figure 7.11). Practically, when a DL is set at what is thought to be its edge-on position, a slight clockwise or anticlockwise tilt, $\pm \Delta\alpha$ around $\alpha = \pi/2$, should immediately lead to the inversion of its contrast.

Finally, once the defects are identified, the calculation of the $|\mathbf{g}\cdot\mathbf{b}|$ and $|\mathbf{g}\cdot\mathbf{b}\otimes\mathbf{u}|$ products for the 12 variants of the {113} RL defect allows us to predict the contrast under which each defect variant should be observed (see Table 7.1).

Using this approach, it is possible to find one precise set of imaging conditions under which parallel and inclined {113} defects can be discriminated [CHE 04]. This is a prerequisite condition for the quantitative analysis of a population of defects. In the following section, we show how this can be used to study the thermal evolution of {113} RL defects during annealing.

7.2.1.3.5. *Statistical analysis of the evolution of population of {113} RL defects during annealing*

Studying the thermal evolution of a population of defects requires that this population is quantified, that is its defect density and its length(or size)-distribution are measured after each annealing step. The defect density we can measure on an image has to be corrected by taking into account the "visibility" of each defect variant, which depends on imaging conditions. The measured sizes or lengths of defects have to be corrected by projection factors that depend on the inclination of these defects with respect to the image plane. Once this is done, the total number of Si atoms contained in the defects can be determined.

In the case of {113} defects, we know that their width is constant (4 nm), independent of their length [CLA 03]. The density, d_{2D}, the mean length, $\langle L \rangle$ and the concentration of trapped atoms within defects per unit area, Nb, can be calculated. We write:

$$\langle L \rangle = \left\langle \frac{L_v}{f_v^{pr}} \right\rangle$$

$$d_{2D} = \sum_v \frac{N_v^g}{f_v^g} \frac{f_S^{pr}}{S} \quad [7.3]$$

$$Nb = d_v^{atom} w \sum_v L_v f_v^{pr}$$

where L_v is the length of a defect variant projection in the image plane, f_v^{pr} is its projection factor (≤ 1), N_v^g is the number of defects visible in the image taken with a given **g**, f_v^g is the correction factor of the defect variant visibility in the image taken with the same **g**, S is the surface of the image under analysis, f_S^{pr} is its projection factor on the (001) plane (≤ 1), d_v^{atom} is the density of Si atoms sitting on the habit plane of the defect and w is the defect width (taken typically equal to 4 nm).

In the example below, populations of {113} RL defects were formed by implanting Si(100) wafers with 2×10^{14} cm^{-2} Si$^+$ ions at 100 keV and annealing at 850°C for times ranging from 150 to 250 s in flowing N$_2$. The populations were analyzed by separating parallel and inclined defects. The total defect density was obtained by summing up the densities of the two types of defect while their mean-length was obtained by averaging their lengths. Then, the effect of the annealing time could be analyzed from the graphs shown in Figure 7.12.

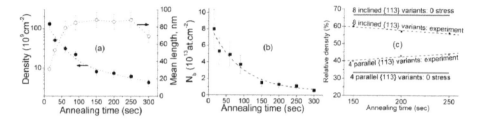

Figure 7.12. *Thermal evolution of a population of {113} RL defects. (a) Mean length (right axis) and density (left axis) as functions of annealing time; (b) number of atoms stored in the population as function of annealing time; (c) relative densities of inclined and parallel {113} RL defects within the population. Comparison between TEM measurements and expectations (no stress), again as a function of annealing time*

During the first 90 s of annealing, the {113} RL defects grow fast in size (Figure 7.12(a), right axis). They reach a maximum average size of about 90 nm after 150 s of annealing. During further annealing, they more or less keep the same average size, but, after 300 s, their size rapidly decreases. In the meantime, the defect density continuously decreases (Figure 7.12(a), left axis). By combining these two measurements, we show that the number of atoms stored in the defect population continuously decreases throughout annealing (Figure 7.12(b)). These graphs clearly show that during annealing the population of defects evolves following a non-conservative Ostwald-type ripening, that is by interchanging the Si atoms they are made of while continuously losing part of these atoms. After 250 s, the density of defects is so low that large defects start to dissolve.

Another interesting piece of information could be extracted from this analysis. As four variants are parallel to the surface while eight are inclined, we would expect that the density of "inclined" defects weight 67% of the total density. Actually, the measurements show that from the beginning of the annealing, these inclined defects are less numerous than expected and that their density decreases faster than the density of parallel defects (Figure 7.12(c)). As a result, the proportion of parallel defects continuously increases during annealing. These parallel defects have their Burgers vector quasi-perpendicular to the wafer surface, that is in a direction where stress relaxation through deformation is possible. On the contrary, the Burgers vectors of inclined defects have large components on the (100) plane, which tend to increase the in-plane compressive stress field generated by the implantation (see section 7.2.1.1). This characteristic renders the nucleation and further growth of the parallel defects "easier" as compared to the inclined defects.

Such analysis show that stress fields may break the crystal symmetry and thus the equivalence between defect and precipitates variants, leading to the preferential growth of the variants that increase the amplitude of these stress fields less than others. The possibility to manipulate populations of defects/precipitates by applying external stress during annealing is being actively investigated at the moment, envisaging many different applications.

7.2.2. Defects of vacancy type

7.2.2.1. Bubbles and voids

During ion implantation, vacancies and interstitials are generated in the same numbers. While most of them recombine during annealing, several process conditions can be used to promote the growth of vacancy-type defects. Interestingly, and in contrast to what is observed in metals, vacancies do not precipitate alone in the form of DLs in Si. In place, there are many reports on the formation of bubbles and voids, that is three-dimensional defects, in Si. Voids, that is "vacuum precipitates", are used in the industry to getter metallic impurities and have been proposed to control boron diffusion during annealing following a vacancy engineering approach [MAR 07, MAR 08]. Voids are, in general, fabricated by implanting a gas (He, H, Ne) followed by high-temperature annealing. During annealing, gas atoms and vacancies coprecipitate and form pressurized bubbles. During further annealing, the gas atoms and/or molecules can exodiffuse from the bubbles toward the surface of the wafer, leaving behind a population of empty voids in place [CER 00].

Voids and bubbles can be visualized using two methods: the Fresnel imaging technique [WIL 96] and off-axis holography (see [TON 93] and Chapter 1). Both methods exploit the phase shift existing between two electron waves passing through materials of different composition/atomic density (mean inner potential or MIP in Chapter 1).

When taking a BF image of a void under strongly defocused conditions (in practice, a defocus of the order of 300–500 nm), the waves emerging from both sides of the void/matrix interface interfere and give rise to Fresnel fringes (one or more pairs of black/white fringes) delineating the interface. When reversing the sign of focus (e.g. going from over to underfocus), the fringe sequence reverses (e.g. from black/white to white/black). As in the case of electron holography, dynamical contrast from the matrix should be minimized by tilting the sample away from exact Bragg conditions. Under such conditions and provided the incident beam is highly coherent, the matrix image is gray and shows low contrast while highly contrasted pairs of Fresnel fringes appear centered on the edges of the voids. The size of every void can be determined with a precision of ±0.5 nm by using the inflection point of the fringe contrast, which exactly defines the interface position [BEN 03]. The ultimate resolution of the technique is principally limited by the size of the objective aperture used to image the sample in BF. Voids (and precipitates) of diameters less than 2 nm can be routinely detected using this technique.

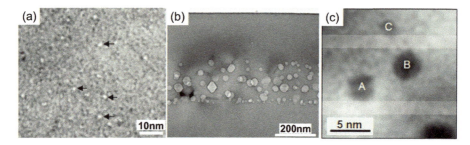

Figure 7.13. *Cross-sectional Fresnel imaging of He bubbles (marked by arrows) (a) and voids (b) (taken from [MAR 07]) obtained in (unannealed) Si implanted with He^+ (a), and after annealing at 1000°C during 1 h (b). Phase image extracted from off-axis holographic image of He bubbles (marked by A, B, C) and formed in iron–chromium alloy after He^+ implantation (taken from [SNO 37])*

Figure 7.13(a) shows a population of He nano-bubbles formed in Si by He implantation. Figure 7.13(b) shows a population of mature, well-facetted voids obtained by long-time annealing of the population of nano-bubbles shown in Figure 7.13(a). From such images, we can extract accurate size distributions. Density and volume fraction measurements can also be obtained from such XTEM images provided the sample thickness is known at every position, which is shown later on in this chapter. Such images have been used to discuss the growth mechanisms of voids in silicon, a still controversial topic.

Off-axis holography can alternatively be used to image the phase differences between the waves passing inside and outside the voids. By forcing the interference between these waves and one reference beam propagating in vacuum, holographic fringes are formed in the image from which the phase can be extracted. The contrast in the "phase image" (see Figure 7.13(c)) is proportional to the MIP (see discussion in Chapter 1) and can be used to measure the size of voids and precipitates. The ultimate resolution of the method is about 0.5–0.6 nm and is mostly limited by the size of the aperture mask used for Fourier filtering. We note here that while both techniques exploit the same physical characteristics (phase difference), electron holography does not provide any advantage in terms of resolution or easiness than the "old" Fresnel technique, as far as the study of populations of voids/precipitates is concerned.

7.2.2.2. *Gas platelets*

As stated in a previous section, hydrogen ion implantation followed by wafer bonding and thermal annealing is used for the transfer of Si films onto oxidized substrates [BRU 99]. This process takes advantage of the precipitation of hydrogen during annealing in the form of quasi-two-dimensional defects filled with overpressurized H_2 gas. These defects exert stress in the layer and, during annealing, they evolve from "platelets" of a few tens of nanometers in diameter to microcracks of a few micrometers in diameter. The possibility for them to elastically interact may give rise to the complete exfoliation of the upper layer [PER 08]. The need to understand the physical mechanisms involved in this complex process has required that the formation and thermal evolution of populations of platelets are quantified, that is the image conditions under which they are observed are selected so that their types, size distribution densities and board-to-board distances can be measured.

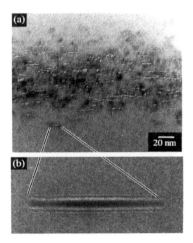

Figure 7.14. *Typical Fresnel contrast arising from platelets. The crystal is set out-of-Bragg conditions and the objective lens is underfocussed by about 400 nm (taken from [GRI 00a])*

Images of platelet populations taken under dynamical and in-focus conditions are, in general, of little interest since the large strain fields generated by every platelet overlap while the defects by themselves are not visible. On the other hand, being precipitates of gas atoms and vacancies, the MIP within these platelets significantly differs from the surrounding Si lattice (as for bubbles). Thus, platelets can be imaged using the Fresnel imaging technique [GRI 00a]. Again, this requires the sample to be set far from dynamical conditions and the image to be taken under strong out-of-focus conditions. An enlarged image of a typical platelet taken under such conditions is shown in Figure 7.14.

In silicon, platelets have been observed to form on the {001} and {111} planes [HEB 07]. Thus, all platelets distribute within two families, that is within seven variants (4 + 3).

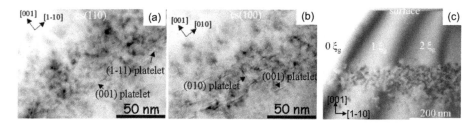

Figure 7.15. *Cross-sectional BF images of H platelets formed after H^+ implantation into (001) Si and annealing at 350°C over 3 min: (a) (110) and (b) (100) cross-sectional out-of-Bragg over focused images (taken from [HEB 07]). The (001), {111} and {010} platelet variants are marked by arrows. (c) (110) cross-sectional BF image taken with $\mathbf{g} = 2\bar{2}0$ in exact Bragg conditions. The specimen thickness in the different zones of the sample is indicated in extinction distances unit, ξ_g*

Figure 7.15 shows a set of typical images used to analyze platelet populations. The different variants are indicated by arrows in Figures 7.15(a) and (b). Platelets parallel to the (001) surface of the wafer can be seen edge-on both in {110} and {100} cross-sectional images. One of the two {010} variants, perpendicular to the surface, can be seen edge-on only in the {100} cross-sectional image. Two of the four {111} variants can be seen edge-on in the {110} cross-sectional image.

Measuring the depth-distribution of all these platelets is of great interest to understand how microcracks will appear. These data are needed to calculate the volume fraction occupied by the platelets, the mean board-to-board distance between platelets and the amount of hydrogen stored within them (provided that the H atomic density within a platelet is known) [GRI 00a, GRI 00b]. However, the measurement of the density of platelets at each depth requires that the thickness of the sample in the analyzed regions is known. This can be obtained by imaging this region under dynamical two-beam conditions. Under such conditions, the thickness fringes can be used to estimate the thickness of the sample [WIL 96]. A "good" beveled sample shows equally spaced fringes (see Figure 7.15(c)). From one bright fringe to the next, the sample thickness has increased by one extinction distance, ξ_g. ξ_g depends on the accelerating voltage of the microscope, on the material (here Si) and on the diffraction vector, **g**, used for imaging. In Figure 7.15(c), $V = 200\,\text{keV}, \mathbf{g} = 2\bar{2}0, s = 0$ and thus $\xi_g = 96\,\text{nm}$. The absolute error that can be made is about $\pm 1/4 \xi_g$. Thus, the relative error in the thickness measurement is minimized when counting platelets in somehow thick regions, provided they are all easily detectable in the corresponding Fresnel image.

The H concentration contained within the platelets distributed at a certain depth and within a band of length, L, and width, w, is given by:

$$c_H = \frac{N_i \sum \pi r_i^2 \ell}{Lwt} c_H^a = \phi c_H^a \qquad [7.4]$$

where r_i is the radius of the platelet i, ℓ is the thickness of a platelet equal to 0.5–1 nm, c_H^a is the atomic density of H atoms within a platelet and equal approximately to $2 \times 10^{23}\,\text{cm}^{-3}$ GRI 00b], N is the number of platelets contained in the band of measured thickness, t and ϕ is the volume fraction occupied by platelets. For platelets parallel to the (001) surface, the average center-to-center distance, and more interestingly the average board-to-board distances $<l>$, in a plane parallel to the (001) surface can be calculated. If we assume that the platelets are randomly and uniformly distributed as parallel flattened cylinders having average radius $<r>$, the ratio $\langle l \rangle / \langle r \rangle$ can be expressed [BAN 72]:

$$\frac{\langle l \rangle}{\langle r \rangle} = \frac{\exp(4\phi)}{2\phi^{1/2}} \Gamma\left(\frac{1}{2}, 4\phi\right) \qquad [7.5]$$

where $\Gamma(1/2, 4\phi) = \int_{4\phi}^{\infty} x^{-1/2} \exp(-x)\,dx$ is a gamma-function.

Figure 7.16 shows an example of the analysis that can be performed following the statistical measurement of the depth-distributions of the seven platelet variants formed in a (001) Si wafer implanted with two different H fluencies, namely a low fluence of $1 \times 10^{16}\,\text{cm}^{-2}$ (Figures 7.16(a)–(c)) and a high fluence of $3 \times 10^{16}\,\text{cm}^{-2}$ (Figures 7.16(d)–(f)), and similarly annealed.

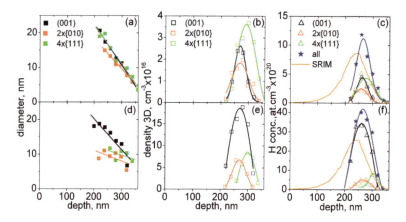

Figure 7.16. Depth distributions of diameters (a–d), 3D densities (b–e) and H concentrations (c–f) for platelets, parallel to the surface ((001) variant), perpendicular to the surface (two {010} variants) and inclined with respect to the surface (four {111} variants), as a function of the implanted fluence. (a), (b) and (c): 1×10^{16} cm^{-2}. (d), (e) and (f): 3×10^{16} cm^{-2}. In (c) and (f), the H profiles predicted by SRIM are superimposed

Some very important information can be extracted from these data. Whatever their type, the platelet size linearly decreases from the surface toward the substrate both in the samples implanted at low and high fluences. The size of the (001) platelets does not depend on the fluence. However, the size of all the other variants decreases when the fluence increases. The {111} platelets distribute deeper in the substrate than the {001} platelets and this effect is more pronounced for the sample implanted with the high fluence. The density of {111} platelets is larger than that of the (001) platelet in the sample implanted at low fluence while it is contrary in the sample implanted with the high fluence. The {010} platelets perpendicular to the surface always have the lowest density. The amount of hydrogen contained within the different platelets follows their density depth-distribution (see Figures 7.16(c) and (f)). Finally, the comparison between the H profile obtained by the Monte Carlo SRIM simulation and the depth-distributions of H contained within all platelets shows that most of the implanted H is contained within the platelets.

These data can be understood by (again) considering the effect of the stress generated by the implantation on the distribution of crystallographic variants and families of platelets. Indeed, this stress increases when the fluence increases (Figure 7.6). Thus, as discussed about {113} RL defects, the nucleation and further growth of platelet variants having a non-zero in-plane component of Burgers vector is less favorable [NAS 05] than that of (001) platelets parallel to the surface. The distribution of the different variants depends on the exact value of this component and on the in-plane stress still present in the substrate. However, stress cannot explain two experimental facts. Following only the same argument, we would expect the {010} platelets perpendicular to the surface not to be stable since they provide Burgers vectors lying in the (001) plane. This is not the case, since they are distributed over the same depths as the (001) platelet variants. Moreover, all the {010} variants have the same or a higher density than all the {111} variants. This suggests that other ingredients must play a role such as reaction barriers.

However, this example shows that the full and independent characterization of defect families and variants is mandatory to approach and eventually understand the difficult interplay between stress and defects in silicon.

7.3. Conclusions

Many of the processing steps used for device manufacturing result in the formation of defects. Identifying these defects, that is determining their nature, Burgers vector and habit plane, is necessary to understand their origin. Conventional techniques such as WBDF and Fresnel imaging can be used for that purpose but new techniques such as GPA of HREM images and electron holography can provide supplementary information. In particular, DFEH is well adapted to detect defects of very small sizes, which cannot be imaged by conventional methods, by evidencing the strain fields they generate. Moreover, defect characteristics must also be known to predict the image contrast rules to which they obey and select the image conditions under which populations of defects can be statistically analyzed. The statistical analysis of defect populations in terms of size distribution and density allows us to study, understand and eventually predict the influence of process conditions onto these defects. The examples shown in this chapter cover most of the defects (extrinsic and intrinsic) that can be found in Si and related compounds after processing.

7.4. Bibliography

[ASO 01] ASPAR B., MORICEAU H., JALAGUIER E., LAGAHE C., SOUBIE A., BIASSE B., PAPON A.M., CLAVERIE A., GRISOLIA J., BENASSAYAG G., LETERTRE F., RAYSSAC O., BARGE T., MALEVILLE C., GHYSELEN B., "The generic nature of the Smart-Cut(R) process for thin film transfer", *Journal of Electronic Material*, vol. 30, pp. 834–840, 2001.

[BAN 72] BANSAL P. P., ARDELL A.J., "Average nearest-neighbor distances between uniformly distributed finite particles", *Metallography*, vol. 5, pp. 97–111, 1972.

[BEN 95a] BENAMARA M., ROCHER A., LAPORTE A., SARRABAYROUSE G., LESCOUZERES L., PEYRE-LAVIGNE A., CLAVERIE A., "Atomic structure of the interfaces between silicon directly bonded wafers", *Proceedings of Material Research Society*, San Francisco, CA, vol. 378, pp. 863–868, 1995.

[BEN 95b] BENAMARA M., "Contribution à l'étude des interfaces de soudure directe dans le silicium", PhD Thesis, University of Toulouse, 1995.

[BEN 03] BEN ASSAYAG G., BONAFOS C., CARRADA M., CLAVERIE A., NORMAND P., TSOUKALAS D., "Transmission electron microscopy measurements of the injection distances in nanocrystal-based memories", *Applied Physics Letters*, vol. 82, pp. 200–202, 2003.

[BIS 08] BISOGNIN G., VANGELISTA S., BRUNO E., "High-resolution X-ray diffraction by end of range defects in self-amorphized Ge", *Materials Science and Engineering B*, vol. 154, pp. 64–67, 2008.

[BON 98] BONAFOS C., MATHIOT D., CLAVERIE A., "Ostwald ripening of end-of-range defects in silicon", *Journal of Applied Physics*, vol. 83, pp. 3008–3017, 1998.

[BON 06] BONINELLI S., CHERKASHIN N., CRISTIANO F., CLAVERIE A., "Evidences of an intermediate rodlike defect during the transformation of {113} defects into dislocation loops", *Applied Physics Letter*, vol. 89, pp. 161904-1-161904-3, 2006.

[BRU 99] BRUEL M., "Separation of silicon wafers by the smart-cut method", *Material Research Innovations*, vol. 3, pp. 9–13, 1999.

[CER 00] CEROFOLINI G.F., CALZOLARI G., CORNI F., FRABBONI S., NOBILI C., OTTAVIANI G., TONINI R., "Thermal desorption spectra from cavities in helium-implanted silicon", *Physical Review B*, vol. 61, pp. 10183–10193, 2000.

[CHE 04] CHERKASHIN N., CALVO P., CRISTIANO F., de MAUDUIT B., COLOMBEAU B., LAMRANI Y., CLAVERIE A., "On the "life" of {113} defects", *Proceedings of Material Research Society*, San Francisco, vol. 810, pp. C3.7.1– C3.7.6, 2004.

[CHE 12] CHERKASHIN N., KONONCHUK O., REBOH S., HYTCH M., "Application of the O-lattice theory for the reconstruction of the high-angle near 90 degrees tilt Si(110)/(001) boundary created by wafer bonding", *Acta Materialia*, vol. 60, no. 3, pp. 1161–1173, 2012.

[CLA 95] CLAVERIE A., LAANAB L., BONAFOS C., "On the relation between dopant anomalous diffusion in Si and end-of-range defects", *Nuclear Instruments and Methods in Physics Research Section B*, vol. 96, no. 1–2, pp. 202–209, 1995.

[CLA 03] CLAVERIE A., COLOMBEAU B., DE MAUDUIT B., BONAFOS C., HEBRAS X., BEN ASSAYAG G., CRISTIANO F., "Extended defects in shallow implants", *Applied Physics A*, vol. 76, pp. 1025–1033, 2003.

[COW 99] COWERN N.E.B., ALQUIER D., OMRI M., CLAVERIE A., NEJIM A., "Transient enhanced diffusion in preamorphized silicon: the role of the surface", *Nuclear Instruments and Methods in Physics Research Section B*, vol. 148, pp. 257–261, 1999.

[DAG 95] EAGLESHAM D.J., STOLK P.A., GOSSMANN H.J., HAYNES T.E., POATE J.M., "Implant damage and transient enhanced diffusion in Si", *Nuclear Instruments and Methods in Physics Research Section B*, vol. 106, pp. 191–197, 1995.

[DE 94] DE MAUDUIT B., LAANAB L., BERGAUD C., FAYE M.M., MARTINEZ A., CLAVERIE A., "Identification of EOR defects due to the regrowth of amorphous layers created by ion bombardment", *Nuclear Instruments and Methods in Physics Research Section*, B vol. 84, pp. 190–194, 1994.

[EDI 75] EDINGTON J.W., *Monographs in Practical Electron Microscopy in Materials Science*, section 3.5, Philips Technical Library, Macmillan (Reprinted by Techbooks in 1976), 1975.

[GRI 00a] GRISOLIA J., BEN ASSAYAG G., CLAVERIE A., ASPAR B., LAGAHE C., LAANAB L., "A transmission electron microscopy quantitative study of the growth kinetics of H platelets in Si", *Applied Physics Letter*, vol. 76, no. 7, pp. 852–854, 2000.

[GRI 00b] GRISOLIA J., Evolution thermique des défauts introduits par implantation ionique d'hélium ou d'hydrogène dans le silicium et le carbure de silicium, PhD Thesis, University of Toulouse, 2000.

[GRO 61] GROVES G.W., KELLY A., "Interstitial dislocation loops in magnesium oxide", *Philosophical Magazine*, vol. 6, p. 1527–1529, 1961.

[HAR 10] HARTMANN J.M., SANCHEZ L., VAN DEN DAELE W., ABBADIE A., BAUD L., TRUCHE R., AUGENDRE E., CLAVELIER L., CHERKASHIN N., HYTCH M., CRISTOLOVEANU S., "Fabrication, structural and electrical properties of compressively strained Ge-on-insulator substrates", *Semiconductor Science Technology*, vol. 25, pp. 075010-1-075010-11, 2010.

[HEB 07] HEBRAS X., NGUYEN P., BOURDELLE K.K., LETERTRE F., CHERKASHIN N., CLAVERIE A., "Comparison of platelet formation in hydrogen and helium-implanted silicon", *Nuclear Instruments and Methods in Physics Research Section B*, vol. 262, pp. 24–28, 2007.

[HŸT 03] HŸTCH M.J., PUTAUX J.-L., PÉNISSON J.-M., "Measurement of the displacement field of dislocations to 0.03 angstrom by electron microscopy", *Nature*, vol. 423, pp. 270–273, 2003.

[KIT 09] KITTLER M., REICHE M., "Dislocations as Active Components in Novel Silicon Devices", *Advanced Engineering Materials*, vol. 11, pp. 249–258, 2009.

[KOF 09a] KOFFEL S., SCHEIBLIN P., CLAVERIE A., BENASSAYAG G., "Amorphization kinetics of germanium during ion implantation", *Journal of Applied Physics*, vol. 105, pp. 013528-1–013528-5, 2009.

[KOF 09b] KOFFEL S., CHERKASHIN N., HOUDELLIER F., HYTCH M.J., BENASSAYAG G., SCHEIBLIN P., CLAVERIE A., "End of range defects in Ge", *Journal of Applied Physics*, vol. 105, pp. 126110-1–126110-3, 2009.

[LAA 95] LAANAB L., BERGAUD C., BONAFOS C., CLAVERIE A., "Variation of end of range density with ion beam energy and the predictions of the "excess interstitials" model", *Nuclear Instruments and Methods in Physics Research Section B*, vol. 96(1–2), pp. 236–240, 1995.

[MAR 07] MARCELOT O., CLAVERIE A., CRISTIANO F., CAYREL F., ALQUIER D., LERCH W., PAUL S., RUBIN L., JAOUEN H., ARMAND C., "Effect of voids-controlled vacancy supersaturations on B diffusion", *Nuclear Instruments and Methods in Physics Research Section B*, vol. 257, pp. 249–252, 2007.

[MAR 08] MARCELOT O., CLAVERIE A., ALQUIER D., CAYREL F., LERCH W., PAUL S., RUBIN L., RAINERI V., GIANNAZZO F., JAOUEN H., "Diffusion and activation of ultra shallow boron implants in silicon in proximity of voids", *Solid State Phenomena*, vol. 131, pp. 357–362, 2008.

[MIC 03] MICLAUS C., GOORSKY M.S., "Strain evolution in hydrogen-implanted silicon", *Journal of Physics D: Applied Physics*, vol. 36, pp. 177–180, 2003.

[NAS 05] NASTASI M., HÖCHBAUER T., LEE J.-K., MISRA A., HIRTH J. P., RIDGWAY M., LAFFORD T., "Nucleation and growth of platelets in hydrogen-ion-implanted silicon", *Applied Physics Letter*, vol. 86, pp. 154102-1–154102-3, 2005.

[PAT 97] PATRIARCHE G., JEANNÈS F., OUDAR J.L., GLAS F., "Structure of the GaAs/InP interface obtained by direct wafer bonding optimised for surface emitting optical devices", *Journal of Applied Physics*, vol. 82, pp. 4892–4903, 1997.

[PER 08] PERSONNIC S., LETERTRE F., TAUZIN A., CHERKASHIN N., CLAVERIE A., FORTUNIER R., KLOCKER H., "Impact of the transient formation of molecular hydrogen on the microcrack nucleation and evolution in H-implanted Si (001)", *Journal of Applied Physics*, vol. 103, pp. 023508-1–023508-9, 2008.

[REB 09] REBOH S., DE MATTOS A.A., BARBOT J.F., DECLEMY A., BEAUFORT M.F., PAPALEO R.M., BERGMANN C.P., FICHTNER P.F.P., "Localized exfoliation versus delamination in H and He coimplanted (001) Si", *Journal of Applied Physics*, vol. 105, pp. 093528-1–093528-6, 2009.

[SNO 06] SNOECK E., MAJIMEL J., RUAULT M.O., HŸTCH M.J., "Characterization of helium bubble size and faceting by electron holography", *Journal of Applied Physics*, vol. 100, pp. 023519-1–023519-5, 2006.

[SOU 06] SOUSBIE N., CAPELLO L., EYMERY J., RIEUTORD F., LAGAHE C., "X-ray scattering study of hydrogen implantation in silicon", *Journal of Applied Physics*, vol. 99, pp. 103509-1–103509-6, 2006.

[TAK 91a] TAKEDA S., "An atomic model of electron-irradiation-induced defects on {113} in Si", *Japan Journal of Applied Physics*, vol. 30, pp. L639–L642, 1991.

[TAK 94] TAKEDA S., KOHYAMA M., IBE K., "Interstitial defects on (113) in Si and Ge – line defect configuration incorporated with a self-interstitial atom chain", *Philosophical Magazine A 70*, pp. 287–312, 1994.

[TON 93] TONOMURA A., *Electron Holography*, Springer-Verlag, Berlin, 1993.

[WIL 96] WILLIAMS D.B., BARRY CARTER C., *Transmission Electron Microscopy*, vol. 1–4, Plenum Press, New York, 1996.

[ZOG 07] ZOGRAPHOS N., ZECHNER C., AVCI I., "Efficient TCAD model for the evolution of interstitial clusters, {311} defects, and dislocation loops in silicon", *Material Research Society Proceedings*, vol. 994, pp. 297–305, 2007.

Chapter 8

In Situ Characterization Methods in Transmission Electron Microscopy

8.1. Introduction

"*In situ* transmission electron microscopy" (*in situ* TEM) is a widely used term that easily enhances an experimental description. As an example, the use of a high-energy electron beam to irradiate a sample is often called an *in situ* TEM technique [KLI 00a, LIU 12a] while the claimed stimuli is in fact the medium used to obtain an image of the sample. One reason could be that *in situ* (from the Latin phrase for "in place" or more commonly "at the right place") is also an international word, easy to understand and exactly pronounced alike by almost every TEM user in the world (unlike TEM that has at least two known forms in Europe: TEM and MET). *In situ* TEM is thus a continuously growing topic. From three occurrences per year in the 1980s (~1% of TEM papers), *in situ* TEM has reached more than 300 occurrences (~7% of TEM papers) per year since 2010[1].

The aim of this chapter is to give some explanations about *in situ* TEM mostly dedicated to the study of devices. It will try to be as exhaustive as possible. However, as a snapshot of a (quickly) growing topic, it will perhaps be obsolete in the next 10 years. This reminder is here to inform the reader that this chapter was written in the middle of 2012 with the results and instrumental breakthroughs of that year, which could perhaps be out of date when it is read.

Chapter written by Aurélien MASSEBOEUF.
1 Citation analysis realized via ISI web of knowledge (c), last consulted on September 1 2012

The first aim of *in situ* is to bring a new dimension to a TEM result. *In situ* techniques are then directly associated with quantitative measure linked to a TEM image. The new dimension being brought by the experimentalist, it is assumed to be perfectly quantified. The analysis that is made, therefore, is a quantification along the new dimension of conventional TEM images. Quantitative imaging often claimed and required for TEM results does not require, for the *in situ* case, complex image formation knowledge or high-level contrast simulation to bring quantification. Nevertheless, a perfect knowledge of the physics at play is absolutely needed to link macroscopic observables to what is happening at the microscopic scale.

The various forms of *in situ* techniques will be listed regarding their physical meaning conjugated to their experimental requirements. *Temperature* can be modified by heating or cooling the sample. For that specific case, numerous commercial solutions are available. *Electromagnetic field* applications are quite easy to produce inside a TEM; first of all, because they are non-local constraints, but also because TEMs are fitted with electromagnetic lenses that can be used for their generation. *Mechanical* traction or compression is an old topic that will be briefly discussed. Focus will preferentially be made on the emergence of micro-electromechanical system (MEMS) integration to produce a truly local and small displacement. *Chemistry* and *light interaction* are lightly introduced as they mostly rely on a microscope modification. Finally, we will add the *current* injection constraint that is by far an integration of almost all the previous topics. This last experimental group is on the one hand a stand-alone topic as current can be used both resistance measurement as well as for magnetic domain wall motion. However, on the other hand, injected current is also the driving parameter for a temperature increase or a MEMS actuation.

The following section is thus dedicated to a presentation of these various forms of *in situ* TEM techniques. The actual and future solutions available for injecting currents into devices are given in the following section. This section will also give a review of commercial and homemade sample holder designs. As samples sometimes need to be adapted for *in situ* experiments, a last section has been dedicated to that specific part.

8.2. *In situ* in a TEM

So to speak, *in situ* TEM has the same approach as tomography or dynamical TEM where a third length dimension or a time scale is, respectively, added to a conventional bidimensional image. Here, the added dimension is not metrological but analytical. Nevertheless, such new dimensions can *a priori* be added using more complex modes of TEM such as spectroscopy or holography. Last but not least, such new dimensions added with *in situ* techniques are additive. For example, one could imagine an experiment simply bringing a third length dimension along with a temperature and a magnetic field scale. Therefore, two more dimensions would be added to the tomogram, giving rise to an extensive quantitative tomography experiment with five quantified dimensions. We discuss in that section the different classes of *in situ* experiment that can be driven inside a TEM.

8.2.1. *Temperature control and irradiation*

One of the first effects of an electron beam interacting with a sample is local heating due to inelastic scattering (even if such heating is in the order of a few degrees [EGE 04]). The temperature thus quickly became a parameter that has to be managed during TEM experiments. It is thus now possible to adjust a sample temperature toward low temperatures as well as higher ones: from around the boiling temperature of helium (~ 10 K) [SUT 11] to the melting point of silicon (1,700 K) [NIS 02].

The major concern in such a case is the thermal conduction between the sample and the cooling or heating point. Cooling capacities rely on a fluid circulation within the sample holder (sometimes with specifications added to the microscope itself). Time is the key to let the whole system adapt to the desired temperature. The important measure here is the temperature as close as possible to the object. This is traditionally realized using an exchangeable thermocouple that is sensitive to the selected temperature range, but is far from the field of view. For high temperatures, even if commercial designs enable to use standard TEM specimens, the temperature is limited to few hundred degrees. Higher temperatures are only possible by directly putting samples onto a filament that will be heated [NIS 02, YON 00] (see Figure 8.1(d)). That configuration only ensures a good heat transfer between the heating source and the sample. Such a design also limits the kind of sample that can be used. Nevertheless, in such cases electron irradiation can be used to locally tune the temperature using a small STEM (scanning TEM) probe. At this time, electron irradiation is mostly used to induce structure change in fragile samples such as carbon-based objects [LIU 12a] or local formation of precipitates within thin foils [KLI 00b]. More complex systems also add an on-chip designed thermometer to perfectly measure the temperature close to the observed part of the sample [HAR 11].

8.2.2. *Electromagnetic field*

As mentioned before, electrons are focused within a TEM using an electromagnetic field. It is therefore easy to use the imaging magnetic field as an applied constraint. Lorentz microscopy [CHA 84] is thus often used in conjunction with magnetic field application. One of the major advantages of electromagnetic fields is their relative homogeneity regarding TEM sample size. The magnetic (or electric) field, for example, is considered as constant within the sample region of the pole piece, or at least on the observation area when designed on the whole TEM sample [KLI 10]. Nevertheless, lots of developments are still being made on TEM holder design or the design of the sample itself [BRI 10] to provide electromagnetic fields in a local approach [TAK 06] and in specific orientations [KLI 10, CUM 08, UHL 03] or in a pulsed form [YI 04].

8.2.3. *Mechanical*

Along with electron irradiation, mechanical stress was perhaps the first *in situ* field of TEM. The first approach is to use a holder applying the strength uniformly at the macroscopic scale [KUB 79] (see Figure 8.1(c)). In such a case, the sample has to be previously designed to enable a thin area for TEM observation as well as an overall design to afford such a mechanical constraint. These *in situ* observations are generally carried out using diffraction, conventional or even high-resolution imaging [OH 09]. Following the general evolution of TEM sample holder design, new forms of strain application have appeared in the last decade. Among others are the use of a moving probe [STA 01] for local indentation or MEMS devices to carefully apply controlled forces on reduced dimension objects such as material covered tips [ISH 10], nanowires [PAN 11a] or nanotubes [MUO 09a]. Such an integration is thus becoming one of the major concerns for *in situ* strain experimentalists, as it is the only known method to associate *in situ* strain application with the emerging methods of TEM such as electron holography or tomography [MID 09]. The moving probe is discussed in section 8.3.2 (see also Figure 8.1(a)).

8.2.4. *Chemistry*

One of the most well known dreams of chemists would be the observation of chemical reactions at the atomic scale. Environmental microscopy is a continuously growing topic that brings some important instrumental developments to the TEM community. The two main classes of environmental TEM (ETEM) are the observation of specimens under a gaseous [YOK 12] or liquid [DEJ 11] atmosphere. The main tools used for these TEM variations are called TEM windows [CRE 08]. They are based on nitride or oxide silicon membranes. A detailed presentation of this will be given in section 8.4.2. The principle in ETEM is to confine the desired atmosphere within a restricted volume of the TEM chamber included between two of these membranes. Nevertheless, such a designed volume is not sufficient for many reactions or is too much fragile. Dedicated microscopes have thus been developed [GAI 02] to offer a full reaction chamber at the objective stage. Here, the chamber is isolated from the rest of the column by using differential apertures.

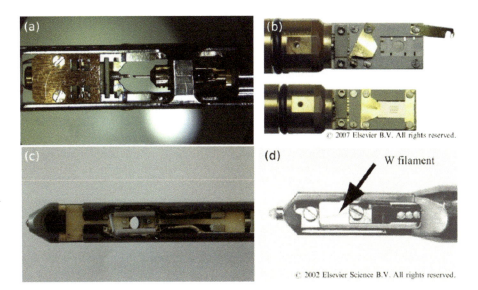

Figure 8.1. *Sample holders for in situ (a) holder with a scanning probe for indentation, biasing or field emission measurement. (b) Custom-made sample holder for multicontacted sample (eight contact pads – courtesy of [KIM 08b]). (c) Sample holder for strain (traction) experiments. (d) Sample holder for high temperature observations (courtesy of [NIS 02])*

8.2.5. *Light*

Light interaction is becoming a timely topic since the new form of microscopy has been discovered such as orbital moments electron beams [VER 10] and plasmons mapping [NEL 07]. Moreover, TEMs designed for ultra-fast [KIM 08a] or dynamical [ZEW 10] microscopy showed the huge interest in bringing light toward a sample. If the column modification is sometimes used for dedicated experiments (as for dynamical processes), the light insertion (or detection) is rather simple using optical fibers [TAN 02]. A challenge remains in conserving preformed light shapes within the TEM column. Like scanning tunneling now commercially available with numerous manufacturers (see also section 8.3.1), it is possible to add a scanning optical probe Scanning Near-Field Optical Microscope (SNOM) within a sample holder [XIA 12]. Sample holders bringing at the same time a probe for electrical biasing or mechanical indentation have also been fabricated [SHI 09].

8.2.6. *Multiple and movable currents*

As a summary of the *in situ* variations presented above, this last point presents the current injection in a general way. Currents are needed in a number of techniques of *in situ* TEM. Magnetic coils, MEMS or thermocouples are, for example, both tuned using currents. Variations inside the panel of current injection are part of the contacts brought near the sample and their versatility (i.e. fixed or movable). Bringing more and more contacts toward the area of interest of the sample theoretically enables a mixing of various forms of *in situ* TEM. In a same way, the possibility of selecting *in situ*, where a current can be applied, is an unaffordable gain regarding the complexity of devices now observed in a TEM. The following sections are thus dedicated to an exploration of the various forms of contacting a sample within a TEM. Initially, the two main families of sample holder are presented (section 8.3) and then the creation of the TEM sample to accept such a contact is explored (section 8.4).

8.3. Biasing in a conventional TEM

As stated above, two approaches may be distinguished in injecting a current within a TEM sample. The historical approach is the opportunity of bringing as many contacts as possible from the outside to the inside of the TEM via the sample holder. Current vacuum feedthrough is well known and widely available, but the connection between the TEM sample and the wire within a sample holder is not unique. We will try to exhaustively describe the available options. A new method was proposed in the late 20th Century with the apparition of STM (scanning tunneling microscope) probes within the sample holders. Last but not least, the combination of these two methods is now emerging. We will give a brief overview of the state of the art of such sample holders.

8.3.1. *Multiple contacts*

At first, an *in situ* TEM experiment aimed to bring wires on the sample under observation via the sample holder. Nowadays, two main characteristics differentiate multi-contact holders: the number and the versatility of the contacts. The more contact there is, the more versatile the sample holder, but the less versatile each contact. The main goal of this section is to give an overview and some examples of various contacting forms that exist in the literature. Nevertheless, an important point in such sample holders is the design of the sample itself, and section 8.4 deals with this subject.

The key point of multi-contact holders (see Figure 8.1(b)) is to offer various constraints on a sample at the same time. One example is the use of currents to heat the sample, to measure the temperature and the resistivity of the sample as well [HAR 11, VER 04]. Such a simple analysis requires at least six independent contacts. When using MEMS inside the TEM to apply local strain, multiple contacts are needed for various parts of the MEMS to be actuated and controlled [ISH 10].

More generally, the number of contacts will determine the number of actuation/measurement couples within the experiment (actuation being multidimensional – indentation is one-dimensional (1D), friction is two-dimensional (2D), etc. – it can rely on multiple contacts as well).

Two methods can be used to connect the sample at the sample holder: clamping or bonding. In the case of clamping, the clamp ensures the mechanical stability as well as the electrical contact [ZHA 05, ZHU 05]. Such a design ensures a perfect reproducibility between sample insertions, but less versatile because all the samples have to be adapted to the clamp. Moreover, the clamping has to be carefully designed to afford a high rate of closing/opening cycles without damaging the sample. The bonding method is more versatile but has to be handled more carefully. Bonding can be made by using wire bonding methods (gold, aluminum and more recently copper) [BRE 10] or simple silver paste. The former is more reliable and conductive but also more traumatic for the samples which may break under the pressure used for bonding, and the latter can sometimes be a poor conductor that can reach up to a few ohms per millimeter.

Finally, the sample has to be integrally designed to accept the contacts. The use of such a sample holder can stand two different types of samples. The most often used type is the free-standing object. The entire direct environment of the object is first designed and the object is then put at the right place for the constraint to be applied. The sample can either be a nanostructure [MUO 09] or a material deposit on a former structure [JAL 12]. In the first case, the sample has to be clamped onto the structure by appropriate welding [PAN 11]. A more complex design is the creation of an entire environment around the existing object. In that case, the object is first deposited onto a "TEM window" that will be lithographied later on [TAN 10, LU 11] (see section 8.4). The main advantage of using such a design is not necessarily to multiply the experimental constraints on an object but to multiply the number of objects under constraint on the same TEM sample (e.g. 10 contacts enable the connexion of five different objects for simple resistivity measurements). First, working transistors based on carbon nanotubes (CNT) were successfully observed using such an approach [KIM 08] (see Figure 8.2(a)).

Some other cases have also been proposed combining a clamping method and a specific sample design, but such methods are somehow limited to a few contacts [TWI 02a]. These experiments were the premise of a new form of holder bringing new forms of contacts.

8.3.2. Movable contacts

A large range of applications appeared with the development of scanning probe microscopies and their piezo-driven mechanics. One of them was the integration of a moving probe inside a TEM holder [IWA 91]. Nevertheless, the low distances applicable with a piezo system only made it difficult to use. It is only at the beginning of the 21st Century that the well-known inertial sliding or "slip-stick" mechanism [POH 87] was practically promoted in an STM probe TEM holder [SVE 03].

The principle was not so much to obtain a coupled STM/TEM image than to use the fine positioning system of a nanometric probe in front of flat surfaces. In such a configuration, the *in situ* experiment was extended to the positioning of the contact itself. The sample is traditionally grounded as the tip brings the bias. Biasing can thus be used to produce field emission from a tip as small as a nanotube [CUM 02]. But the main interest of using such a tip is to inject a current in a specific local area of a device under study. P-n junctions can thus be precisely studied in terms of structure and electrical properties [HAN 08, PAR 10] as well as localized transport measurements [CHI 08], charged memories study [LIU 12b, CHO 11] or various carbon-based material properties [HUA 10, WAN 06, GOL 12]. The ideal observation was the study of a working device using all the sensitivity of TEM techniques. The first experiments on MOSFET devices gave a perfect knowledge of electrostatic potential within the conduction channel [IKA 12] (see Figure 8.2(b)).

The sample design is rather simple as it only needs a free half space to let the probe to come to the surface of interest. FIB processes (described in Chapter 9) are thus perfectly suited to such an analysis. The only precaution is that the sample is effectively grounded.

The TEM holders are still developed to offer more and more interacting forms within the sample regarding *in situ* probing capacities. We can find two moving probes on the same sample holder [MUR 06, KAW 11] or a combination of a probe with a laser injection system [SHI 09]. Manufacturers now propose the combination of a moving probe and multi-contacts within the same sample holders. The four-point measurement method traditionally used in the micro-electronic industry within a TEM sample holder might be accessible in the near future.

8.3.3. Comparison

The TEM experimentalist always has to bear in mind the first *in situ* mechanism presented in this chapter which is electron irradiation. It is worth noting that the few experiments on working devices have pointed out the huge effect of the imaging electron beam on the functions of the devices [IKA 12, KIM 05] (see also Figure 8.2).

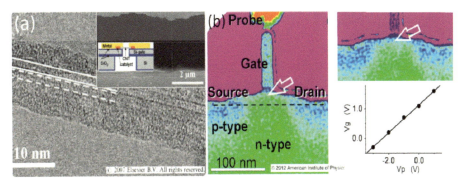

Figure 8.2. *Working devices analyzed in TEM. (a) Field effect transistors (FET) based on CNT bundles mounted in TEM with a multicontact holder. Observation of CNT during FET function finally showed a huge influence of the electron beam irradiation. Inset is showing SEM image of the suspended CNT acting as electron channel (courtesy of [KIM 08]). (b) MOSFET electric potential analysis using a moving probe sample holder. Electron holography is used to reveal the electric potential within the MOSFET channel at 0 (left image) and 2 V (right zoomed image). As electron irradiation is perturbing the device at work (a positive gate tension (V_g) is found for a 0 V (V_p) probing), the authors have taken into account its influence to get quantitative measurement of the gate work function (courtesy of [IKA 12])*

The two presented methods are constantly developed and no one can say today if one will be the standard in a few years. One major consideration could be that the multi-contact holder is less complex to produce as it does not need any complex materials such as piezoelectric or probe holders. Moreover their design is more dramatically important with respect to the available samples. Thus, lots of teams have fabricated their own sample holders, often starting from an originally manufactured simple one. Some teams are still developing some high level moving probe systems [SIR 12]. For specific analysis or money saving, it is sometimes better to produce a homemade sample holder as it will offer without any doubt the versatility expected by the TEM user. Nevertheless, lots of TEM experimentalists might be put off by TEM complexity, from grounding/charging precautions toward ultra-high vacuum problems. We have to add the fact that it is by far faster and easier to get commercially holders for conventional *in situ* operations as a manufactured system comes with all the electronic and software assistance.

The last point that will be discussed here is therefore a manufacturing limitation. Far from wishing to promote any of the TEM manufacturers, it is important to note that only Jeol Ltd. is using side entry sample holders of more that 10 mm in diameter. Such a size gives rise to a lot of space to design complex structures. In fact, almost all the high level developments in sample holders presented in that part have been made with that manufacturer's equipment.

8.4. Sample design

Most of the samples studied in *in situ* experiment are nanostructures. Due to their low dimensionality and uniformity, nanowires (nanotubes) and nanoparticles are well suited to pass a current or to mount on a probe. Nevertheless new TEM sample preparation techniques as well as micro-electronic traditional processes can now be merged to provide complex structures suitable for TEM *in situ* observation. We will see in this section some of the most common tools that can be used to easily design a TEM sample that can be fitted within the sample holders described above. Focused ion beam and nanolithography are now common techniques used to provide thin connected samples. One can start from the bulk specimen where thin foils or devices are located or one can design a dedicated environment around an already TEM compliant sample.

It is worth noting here that common TEM samples can be used for *in situ* techniques such as temperature, strain or magnetic studies. Regarding the moving probes presented above, they can easily be produced regarding the usual processes used for STM tips [IBE 90].

8.4.1. *Focused ion beam*

This section will not deal with traditional sample preparation techniques that can be used to prepare conventional TEM lamella that are now commonly used in the TEM field (and that are extensively examined in Chapter 9). This section is more dedicated to an overview of unusual but simple methods that can lead to well-adapted TEM samples for *in situ* TEM.

The old-fashioned H-bar method [CAS 11] (see also section 9.6) was progressively abandoned in favor of the lift-out method that is by far more convenient for most applications. However, its extremely robustness (the TEM sample belongs to the micrometric scale) is a perfect asset regarding the moving probe system presented above. It offers a flat surface, with a small thin area that can easily be reached by a moving probe within the TEM. Moreover, several H-bar samples can be etched within the same sample that can lead to a high density of TEM windows for destructive indentation experiment [CHI 08]. Last but not least, such a geometry can also be used as a nanowire to pass current in the case of a thin foil (see Figure 8.3.c). A less-known technique is the plane view preparation using FIB (which could be referred to as the U-shape with respect to the H-bar denomination – see Figure 8.3b). Such a technique can easily be used in combination with a few lithographic processes (which can also be done using the FIB) to make a well-suited geometry for strain experiments and/or current injections in nanostructures [BRI 10].

We can also start from a bulk material (small enough to enter a TEM) and use its size to contact it easily within a TEM holder. The FIB is used there to produce a small visible area within the bulk sample in which the probed process occurs [TWI 02b]. FIB can also be used to prepare a sharp surface that is used as substrate for material deposition afterward, and then finally used to obtain one (or more) sample at a specific location that will be used for *in situ* measurements [LIU 12b] (see Figure 8.3d).

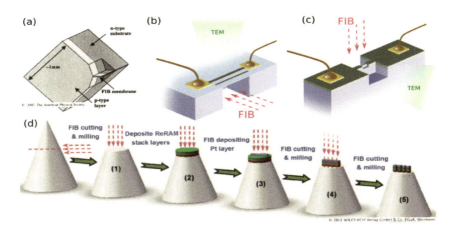

Figure 8.3. *FIB use for in situ samples. (a) TEM lamella made within bulk p-n junction sample using FIB (courtesy of [TWI 02b]). (b) Lithographied plane view sample produced by FIB. (c) Cross-sectional sample used as a nanowire for passing current. (d) FIB preparation of substrate and TEM sample preparation after material deposition (courtesy of [LIU 12b])*

8.4.2. *TEM windows*

In addition to traditional carbon foils deposited on a copper grid, nano-objects are often dispersed on silicon nitride (or oxide) membranes [GRA 04]. This solution, supported by a large commercial offer, is by far more convenient for a lot of *in situ* TEM techniques. First, it proposes a flat surface supported by a relatively thick substrate (few hundred micrometers) that is easy to handle and can be used for material deposition using traditional growing processes (with a few precautions such as protecting the viewing hole from differential pressures). Moreover, the insulating behavior of nitride ensures a perfect isolation of the contacting pads that it supports.

Even with the large panel of commercial offers, it is sometimes better to produce one's own membranes. The traditional TEM window presents small lateral dimensions that do not fit the lithography requirements (sample manipulation and resin deposit are the most important points). Moreover, the instrumental need for such fabrication is very simple and is often found in any cleanroom facility. The main idea is to use a Chemical Vapor Deposition (CVD) grown nitride (resp. thermal oxide) grown on each surface of a thin (less than 300 µm to fit common sample holders) wafer that will be lithographed on one side to create the opening holes. A simple wet (resp. dry) etching is then necessary to remove the silicon toward the opposite nitride (resp. oxide) side. Membrane thickness can be decreased down to a few tenths of a nanometer. UV lithography can then be used to design contact pads and e-beam lithography to connect the sample to the pads. Complex designs (as complex as MEMS) can then be drawn with such technology to reach the concept of "in-TEM micro-laboratory" [HAR 11, VER 04, JAL 12, KAL 12]. In such a concept, the sample is no longer the only object under study, but both the object and its direct environment. Figure 8.4 shows two designs of objects connected to a sample holder through the use of a silicon nitride membrane.

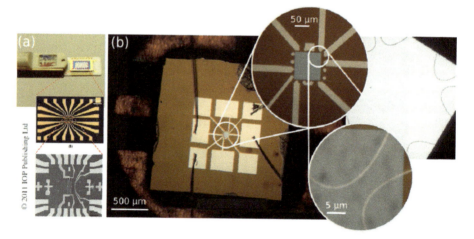

Figure 8.4. *Silicon nitride windows. (a) TEM thermal measurement platform used for nanowires under varying temperature analysis. Black areas in the middle of the bottom image are holes within the nitride membrane to let the electron passing through (courtesy of [HAR 11]). (b) Permalloy stripes (400-nm width) grown on silicon membrane and connected to a sample holder with micro-bonding (four pads connected). A TEM view of the sample (top right) is presented along with optical snapshots*

8.5. Conclusions

In situ TEM is a really rich topic and can still be viewed as a non-traditional microscopy, as it involves a lot of different processes. Mastering all the required processes may be long and tedious, which is why it might sometimes be useful to proceed via the commercially available steps when they exist (sample holders, sample supports, contacting). The TEM experimentalist has to keep in mind that a homemade overall process is more dedicated to the research than a commercial process. Still, *in situ* microscopy is quickly growing mainly due to a lot of description of instrumental and processes developments. As stated previously this chapter might not be up to date for a long time.

8.6. Bibliography

[BRE 10] BREACH C.D., WULFF F.W., "A brief review of selected aspects of the materials science of ball bonding", *Microelectronics Reliability*, vol. 50, no. 1, pp. 1–20, January 2010, available at http://www.sciencedirect.com/science/article/pii/S0026271409003126.

[BRI 10] BRINTLINGER T., LIM S.-H., BALOCH K.H., ALEXANDER P., QI Y., BARRY J., MELNGAILIS J., SALAMANCA-RIBA L., TAKEUCHI I., Cumings J., "In situ observation of reversible nanomagnetic dwitching induced by electric fields", *Nano Letters*, vol. 10, no. 4, pp. 1219–1223, 2010, available at http://dx.doi.org/10.1021/nl9036406.

[CAS 11] CASTANY P., LEGROS M., "Preparation of H-bar cross-sectional specimen for in situ TEM straining experiments: a FIB-based method applied to a nitrided Ti–6Al–4V Alloy", *Materials Science and Engineering: A*, vol. 528, no. 3, pp. 1367–1371, 25 Janurary 2011, available at http://www.sciencedirect.com/science/article/pii/S0921509310011901.

[CHA 84] CHAPMAN J.N., "The investigation of magnetic domain structures in thin foils by electron microscopy", *Journal of Physics D: Applied Physics*, vol. 17, pp. 623–647, 1984.

[CHI 08] CHIARAMONTI A.N., THOMPSON L.J., EGELHOFF W.F., KABIUS B.C., PETFORD-LONG. A.K., "In situ TEM studies of local transport and structure in nanoscale multilayer films", *Ultramicroscopy*, vol. 108, no. 12, pp. 1529–1535, November 2008, available at http://www.sciencedirect.com/science/article/pii/S0304399108000624.

[CHO 11] CHOI S.-J., PARK G.-S., KIM K.-H., CHO S., YANG W.-Y., LI X.-S., MOON J.-H., LEE K.-J., KIM K., "In situ observation of voltage-induced multilevel resistive switching in solid electrolyte memory", *Advanced Materials*, vol. 23, no. 29, pp. 3272–3277, 2011, available at http://onlinelibrary.wiley.com/doi/10.1002/adma.201100507/abstract.

[CRE 08] CREEMER, J.F., HELVEG S., HOVELING G.H., ULLMANN S., MOLENBROEK A.M., SARRO P.M., ZANDBERGEN H.W., "Atomic-scale electron microscopy at ambient pressure", *Ultramicroscopy*, vol. 108, no. 9, pp. 993–998, August 2008, available at http://www.sciencedirect.com/science/article/pii/S0304399108000594.

[CUM 02] CUMINGS J., ZETTL A., MCCARTNEY M.R., SPENCE J.C.H., "Electron holography of field-emitting carbon nanotubes", *Physical Review Letters*, vol. 88, no. 5, p. 056804, 18 January 2002, available at http://link.aps.org/doi/10.1103/PhysRevLett.88.056804.

[CUM 08] CUMINGS, J., OLSSON E., PETFORD-LONG A.K., ZHU Y., "Electric and magnetic phenomena studied by in situ transmission electron microscopy", *Mrs Bulletin*, vol. 33, no. 2, pp. 101–106, February 2008.

[DE 11] DE JONGE N., ROSS F.M., "Electron microscopy of specimens in liquid", *Nature Nanotechnology*, vol. 6, no. 11, pp. 695–704, 2011, available at http://www.nature.com/nnano/journal/v6/n11/abs/nnano.2011.161.html.

[EGE 04] EGERTON R.F., LI P., MALAC M., "Radiation damage in the TEM and SEM", *Micron*, vol. 35, no. 6, pp. 399–409, August 2004, available at http://www.sciencedirect.com/science/article/pii/S0968432804000381.

[GAI 02] GAI P.L., "Developments in in situ environmental cell high-resolution electron microscopy and applications to catalysis", *Topics in Catalysis*, vol. 21, no. 4, pp. 161–173, 1 December 2002, available at http://dx.doi.org/10.1023/A:1021333310817.

[GOL 12] GOLBERG D., COSTA P.M.F.J., WANG M.-S., WEI X., TANG D.-M., XU Z., HUANG Y., et al., "Nanomaterial engineering and property studies in a transmission electron microscope", *Advanced Materials*, vol. 24, no. 2, pp. 177–194, 2012, available at http://onlinelibrary.wiley.com/doi/10.1002/adma.201102579/abstract.

[GRA 04] GRANT, A.W., HU Q.-H., KASEMO B., "Transmission electron microscopy windows for nanofabricated structures", *Nanotechnology*, vol. 15, no. 9, pp. 1175–1181, 1 September 2004, available at http://iopscience.iop.org/0957-4484/15/9/012.

[HAN 08] HAN M.G., SMITH D.J., MCCARTNEY M.R., "In situ electron holographic analysis of biased Si N[sup +]-p junctions", *Applied Physics Letters*, vol. 92, no. 14, p. 143502, 2008, available at http://link.aip.org/link/APPLAB/v92/i14/p143502/s1&Agg=doi.

[HAR 11] HARRIS C.T., MARTINEZ J.A., SHANER E.A., HUANG J.Y., SWARTZENTRUBER B.S., SULLIVAN J.P., CHEN G., "Fabrication of a nanostructure thermal property measurement platform", *Nanotechnology*, vol. 22, no. 27, p. 275308, 8 July 2011, available at http://iopscience.iop.org/0957-4484/22/27/275308.

[HUA 10] HUANG J.Y., QI L., LI J., "In situ imaging of layer-by-layer sublimation of suspended graphene", *Nano Research*, vol. 3, no. 1, pp. 43–50, 5 March 2010, available at http://link.springer.com/article/10.1007/s12274-010-1006-4?null.

[IBE 90] IBE J.P., BEY Jr. P.P., BRANDOW S.L., BRIZZOLARA R.A., BURNHAM N.A., DILELLA D.P., LEE K.P., MARRIAN C.R.K., COLTON R.J., "On the electrochemical etching of tips for scanning tunneling microscopy", *Journal of Vacuum Science & Technology A: Vacuum, Surfaces, and Films*, vol. 8, no. 4, pp. 3570–3575, 1990, available at http://link.aip.org/link/?JVA/8/3570/1.

[IKA 12] IKARASHI N., TAKEDA H., YAKO K., HANE M., "In-situ electron holography of surface potential response to gate voltage application in a sub-30-nm gate-length metal-oxide-semiconductor field-effect transistor", *Applied Physics Letters*, vol. 100, no. 14, p. 143508, 2012, available at http://link.aip.org/link/APPLAB/v100/i14/p143508/s1&Agg=doi.

[ISH 10] ISHIDA T. NAKAJIMA Y., KAKUSHIMA K., MITA M., TOSHIYOSHI H., FUJITA H., "Design and fabrication of MEMS-controlled probes for studying the nano-interface under *in situ* TEM observation", *Journal of Micromechanics and Microengineering*, vol. 20, no. 7, p. 075011, 1 July 2010, available at http://iopscience.iop.org/0960-1317/20/7/075011.

[IWA 91] IWATSUKI M., MUROOKA K., KITAMURA S.-I., TAKAYANAGI K., HARADA Y., "Scanning tunneling microscope (STM) for conventional transmission electron microscope (TEM)", *Journal of Electron Microscopy*, vol. 40, no. 1, pp. 48–53, 1 February 1991, available at http://jmicro.oxfordjournals.org/content/40/1/48.

[JAL 12] JALABERT L., SATO T., TADASHI I., FUJITA H., CHALOPIN Y., VOLZ S., "Ballistic thermal conductance of a lab-in-a-TEM made Si nanojunction", *Nano Letters*, 18 September 2012, available at http://pubs.acs.org/doi/abs/10.1021/nl302379f.

[KAL 12] KALLESØE C., WEN C.Y., BOOTH T.J., HANSEN O., BØGGILD P., ROSS F.M., MØLHAVE K., "In situ TEM creation and electrical characterization of nanowire devices", *Nano Letters*, vol. 12, no. 6, pp. 2965–2970, 13 June 2012, available at http://dx.doi.org/10.1021/nl300704u.

[KAW 11] KAWAMOTO N., WANG M.-S., WEI X., TANG D.-M., MURAKAMI Y., SHINDO D., MITOME M., GOLBERG D., "Local temperature measurements on nanoscale materials using a movable nanothermocouple assembled in a transmission electron microscope", *Nanotechnology*, vol. 22, no. 48, p. 485707, 2 December 2011, available at http://iopscience.iop.org/0957-4484/22/48/485707.

[KIM 08a] KIM J.S., LAGRANGE T., REED B.W., TAHERI M.L., ARMSTRONG M.R., KING W.E., BROWNING N.D., CAMPBELL G.H., "Imaging of transient structures using nanosecond in situ TEM", *Science*, vol. 321, no. 5895, pp. 1472–1475. 12 September 2008, available at http://www.sciencemag.org/content/321/5895/1472.

[KIM 08b] KIM T., KIM S., OLSON E., ZUO J.-M., "In situ measurements and transmission electron microscopy of carbon nanotube field-effect transistors", *Ultramicroscopy*, vol. 108, no. 7, pp. 613–618, June 2008, available at http://www.sciencedirect.com/science/article/pii/S0304399107002227.

[KIM 05] KIM T., ZUO J.M., OLSON E.A., PETROV I., "Imaging suspended carbon nanotubes in field-effect transistors configured with microfabricated slits for transmission electron microscopy", *Applied Physics Letters*, vol. 87, no. 17, pp.173108–173108–3, 18 October 2005, available at http://apl.aip.org/resource/1/applab/v87/i17/p173108_s1.

[KLI 00] KLIMENKOV M., MATZ W., VON BORANY J., "In situ observation of electron-beam-induced ripening of Ge clusters in thin SiO2 layers", *Nuclear Instruments & Methods in Physics Research Section B-Beam Interactions with Materials and Atoms*, vol. 168, no. 3, pp. 367–374, July 2000.

[KLI 10] KLING J., TAN X., JO W., KLEEBE H.-J., FUESS H., RÖDEL J., "In situ transmission electron microscopy of electric field-triggered reversible domain formation in Bi-based lead-free Piezoceramics", *Journal of the American Ceramic Society*, vol. 93, no. 9, pp. 2452–2455, 2010, available at http://onlinelibrary.wiley.com/doi/10.1111/j.1551-2916.2010.03778.x/abstract.

[KUB 79] KUBIN L.P., LOUCHET F., "Analysis of softening in the Fe C system from in situ and conventional experiments-I. In situ experiments", *Acta Metallurgica*, vol. 27, no. 3, pp, 337–342, March 1979, available at http://www.sciencedirect.com/science/article/pii/0001616079900269.

[LIU 12a] LIU J.-W., XU J., NI Y., FAN F.-J., ZHANG C.L., YU S.H., "A family of carbon-based nanocomposite tubular structures created by in situ electron beam irradiation", *ACS Nano*, vol. 6, no. 5, pp. 4500–4507, 22 May 2012, available at http://dx.doi.org/10.1021/nn301310m.

[LIU 12b] LIU Q., SUN J., LV H., LONG S., YIN K., WAN N., LI Y., SUN L., LIU M., "Real-time observation on dynamic growth/dissolution of conductive filaments in oxide-electrolyte-based ReRAM", *Advanced Materials*, vol. 24, no.14, pp. 1844–1849, 2010, available at http://onlinelibrary.wiley.com/doi/10.1002/adma.201104104/abstract.

[LU 11] LU C. WU W.W., OUYANG H., LIN Y.-C., HUANG Y., WANG C.-W., WU Z.W., HUANG C.-W., CHEN L.J., TU K. N., "The influence of surface oxide on the growth of metal/semiconductor nanowires", *Nano Letters*, vol. 11, no. 7, pp. 2753–2758, 13 July 2011, available at http://dx.doi.org/10.1021/nl201037m.

[MID 09] MIDGLEY. A., DUNIN-BORKOWSKI R.E., "Electron tomography and holography in materials science", *Nature Materials*, vol. 8, no. 4, pp. 271–280, 2009, available at http://dx.doi.org/10.1038/nmat2406.

[MOU 09] MUOTH M., GRAMM F., ASAKA K., DURRER L., HELBLING T., ROMAN C., LEE S.-W., HIEROLD C., "Tilted-view transmission electron microscopy-access for chirality assignment to carbon nanotubes integrated in MEMS", *Procedia Chemistry*, vol, 1, no. 1, pp, 601–604, September 2009, available at http://www.sciencedirect.com/science/article/B983C-4X49BRW-5C/2/969746c8e1c0e073e863528dc87c5853.

[MUR 06] MURAKAMI Y., KAWAMOTO N., SHINDO D., ISHIKAWA I., DEGUCHI S., YAMAZAKI K., INOUE M., KONDO Y., SUGANUMA K., "Simultaneous measurements of conductivity and magnetism by using microprobes and electron holography", *Applied Physics Letters*, vol. 88, no. 22, p. 223103, 2006 available at http://link.aip.org/link/APPLAB/v88/i22/p223103/s1&Agg=doi.

[NEL 07] NELAYAH J., KOCIAK M., STEPHAN O., DE ABAJO F.J.G., TENCE M., HENRARD L., TAVERNA D., PASTORIZA-SANTOS I., LIZ-MARZAN L.M., COLLIEX C., "Mapping surface plasmons on a single metallic nanoparticle", *Nature Physics*, vol. 3, no. 5, pp. 348–353, 2007, available at http://dx.doi.org/10.1038/nphys575.

[NIS 02] NISHIZAWA, H. HORI F., OSHIMA R., "In-situ HRTEM observation of the meltingcrystallization process of silicon", *Journal of Crystal Growth*, vol. 236, nos.1–3, pp. 51–58, 2002, available at http://www.sciencedirect.com/science/article/pii/S002202480102156X.

[OH 09] OH S.H. LEGROS M., KIENER D., DEHM G., "In situ observation of dislocation nucleation and escape in a submicrometre aluminium single crystal", *Nature Materials*, vol. 8, no, 2, pp. 95–100, 2009, available at http://www.nature.com/nmat/journal/v8/n2/abs/nmat2370.html.

[PAN 11] PANT B., ALLEN B.L., ZHU T., GALL K., PIERRON O.N., "A versatile microelectromechanical system for nanomechanical testing", *Applied Physics Letters*, vol. 98, no. 5, p. 053506, 2011 available at http://link.aip.org/link/APPLAB/v98/i5/p053506/s1&Agg=doi.

[PAR 10] PARK S., KIM M.J., LOURIE O., "Direct two-dimensional electrical measurement using point probing for doping area identification of nanodevice in TEM", *NANO*, vol. 05, no. 01, p. 61, 2010 available at http://www.worldscientific.com/doi/abs/10.1142/S1793292010001810?journalCode=nano.

[POH 87] POHL D. W., "Dynamic piezoelectric translation devices", *Review of Scientific Instruments*, vol. 58, no. 1, p. 54, 1987, available at http://link.aip.org/link/RSINAK/v58/i1/p54/s1&Agg=doi.

[SHI 09] SHINDO D., TAKAHASHI K., MURAKAMI Y., YAMAZAKI K., DEGUCHI S., SUGA H., KONDO Y., "Development of a multifunctional TEM specimen holder equipped with a piezodriving probe and a laser irradiation port", *Journal of Electron Microscopy*, vol. 58 no. 4, pp. 245–249, 1 August 2009, available at http://jmicro.oxfordjournals.org/content/58/4/245.

[SIR 12] SIRIA A., BAROIS T., VILELLA K., PERISANU S., AYARI A., GUILLOT D., PURCELL, S.T., PONCHARAL P., "Electron fluctuation induced resonance broadening in nano electromechanical Systems: the origin of shear force in vacuum", *Nano Letters*, vol. 12, no. 7, pp. 3551–3556, 11 July 2012, available at http://dx.doi.org/10.1021/nl301618p.

[STA 01] STACH E.A., FREEMAN T., MINOR A.M., OWEN D.K., CUMINGS J., WALL M.A., CHRASKA T., *et al.*, "Development of a nanoindenter for in situ transmission electron microscopy", *Microscopy and Microanalysis*, vol. 7, no. 06, pp. 507–517, 2001.

[SUT 11] SUTTER A., SUTTER P.W., UCCELLI E., FONTCUBERTA I MORRAL A., "Supercooling of nanoscale Ga drops with controlled impurity levels", *Physical Review B*, vol. 84, no. 19, p. 193303, 23 November 2011, available at http://link.aps.org/doi/10.1103/PhysRevB.84.193303.

[SVE 03] SVENSSON K., JOMPOL Y., OLIN H., OLSSON E., "Compact design of a transmission electron microscope-scanning tunneling microscope holder with three-dimensional coarse motion", *Review of Scientific Instruments*, vol. 74, no. 11, pp. 4945–4947, 1 November 2003, available at http://rsi.aip.org/resource/1/rsinak/v74/i11/p4945_s1.

[TAK 06] TAKEGUCHI M., SHIMOJO M., CHE R., FURUYA K., "Fabrication of a nano-magnet on a Piezo-driven tip in a TEM sample holder", *Journal of Materials Science*, vol. 41, no. 9, pp. 2627–2630, 2006 available at http://dx.doi.org/10.1007/s10853-006-7825-8.

[TAN 02] TANABE. MUTO S. and TOHTAKE S., "Development of new TEM specimen holder for cathodoluminescence detection", *Journal of Electron Microscopy*, vol. 51, no. 5, pp. 311–313, 22 October 2002, available at http://jmicro.oxfordjournals.org/content/51/5/311.

[TAN 10] TANG J., WANG C.-Y., XIU F., HONG A.J., CHEN S., WANG M., ZENG C., et al., "Single-crystalline Ni2Ge/Ge/Ni2Ge nanowire heterostructure transistors", *Nanotechnology*, vol. 21, no. 50, p. 505704, 17 December 2010, available at http://iopscience.iop.org/0957-4484/21/50/505704.

[TWI 02] TWITCHETT A.C., DUNIN-BORKOWSKI R.E., MIDGLEY P.A., "Quantitative electron holography of biased semiconductor devices", *Physical Review Letters*, vol. 88, no. 23, p. 238302, 2002, available at http://link.aps.org/doi/10.1103/PhysRevLett.88.238302.

[UHL 03] UHLIG T., HEUMANN M., ZWECK J., "Development of a specimen holder for in situ generation of pure in-plane magnetic fields in a transmission electron microscope", *Ultramicroscopy*, vol. 94, nos. 3–4, pp. 193–196, 2003, available at http://www.sciencedirect.com/science/article/pii/S0304399102002644.

[VER 04] VERHEIJEN M.A., DONKERS J.J.T.M., THOMASSEN J.F.P., VAN DEN BROEK J.J., VAN DER RIJT R.A.F., DONA M.J.J., SMIT C.M., "Transmission electron microscopy specimen holder for simultaneous in situ heating and electrical resistance measurements", *Review of Scientific Instruments*, vol. 75, no. 2, pp. 426–429, 1 February 2004, available at http://rsi.aip.org/resource/1/rsinak/v75/i2/p426_s1.

[VER 10] VERBEECK J., TIAN H., SCHATTSCHNEIDER P., "Production and application of electron vortex beams", *Nature*, vol. 467, p. 301, 2010.

[WAN 06] WANG Y.G., WANG T.H., LIN X.W., DRAVID V.P., "Ohmic contact junction of carbon nanotubes fabricated by in situ electron beam deposition", *Nanotechnology*, vol. 17, no. 24, pp. 6011–6015, 28 December 2006, available at http://iopscience.iop.org/0957-4484/17/24/018.

[XIA 12] XIANG B., HWANG D.J., IN J.B., RYU S.-G., YOO J.-H., DUBON O., MINOR A.M., GRIGOROPOULOS C.P., "In situ TEM near-field optical probing of nanoscale silicon crystallization", *Nano Letter,* 2012, available at http://dx.doi.org/10.1021/nl3007352.

[YI 04] YI G., NICHOLSON W.A.P., LIM C.K., CHAPMAN J.N, MCVITIE S., WILKINSON C.D.W., "A new design of specimen stage for in situ magnetising experiments in the transmission electron microscope", *Ultramicroscopy*, vol. 99, no. 1, pp. 65–72, February 2004 available at http://www.sciencedirect.com/ science/article/pii/ S0304399103001487.

[YOK 12] YOKOSAWA C.P., ALAN T., PANDRAUD G., DAM B., ZANDBERGEN H., "In-situ TEM on (de)hydrogenation of Pd at 0.5–4.5 bar hydrogen pressure and 20–400°C", *Ultramicroscopy*, vol. 112, no. 1, pp. 47–52, January 2012, available at http://www.sciencedirect.com/science/article/pii/S0304399111002555.

[YON 12] YONEZAWA ARAI S., TAKEUCHI H., KAMINO T., KURODA K., "Preparation of naked silver nanoparticles in a TEM column and direct in situ observation of their structural changes at high temperature", *Chemical Physics Letters*, vol. 537, pp. 65–68., 1 June 2012 available at http://www.sciencedirect.com/science/article/pii/ S0009261412004381.

[ZEW 10] ZEWAIL A.H., "Four-dimensional electron microscopy", *Science*, vol. 328, no. 5975, pp. 187–193, 2010 available at http://www.sciencemag.org/cgi/content/abstract/328/5975/187.

[ZHA 05] ZHANG M., OLSON E.A., TWESTEN R.D., WEN J.G., ALLEN L.H., ROBERTSON I.M., PETROV I., "In situ transmission electron microscopy studies enabled by microelectromechanical system technology", *Journal of Materials Research*, vol. 20, no. 07, pp. 1802–1807, 2005.

[ZHU 05] ZHU Y., ESPINOSA H.D., "An electromechanical material testing system for in situ electron microscopy and applications", *Proceedings of the National Academy of Sciences of the United States of America*, vol. 102, no. 41, pp. 14503–14508, 11 October 2005, available at http://www.pnas.org/content/102/41/14503.

Chapter 9

Specimen Preparation for Semiconductor Analysis

Specimen preparation is absolutely the key for quantitative transmission electron microscopy (TEM) analysis; modern transmission electron microscopes are very powerful and stable and, generally speaking, the accuracy of the analysis is now mainly limited by the quality of the specimen. It is a lot easier to learn how to make a good sample, than to perform the complicated analysis that is required for non-perfect TEM lamella.

One of the key differences between specimen preparation for semiconductor research and other fields is that in nearly all cases, nanometer-scale site specificity is required. In addition, such specimens can be made up of many different types of materials in this nanoscaled volume. As a result, the focused ion beam (FIB) milling tool is now indispensible. The FIB is an extremely versatile tool that uses an ion beam, metal deposition, micromanipulators and secondary electron detectors for cutting, sticking, modifying, connecting and imaging materials with nm-scale accuracy and resolution. There already exist many excellent books that present the FIB in detail. Here, we will only discuss the background of how specimens can be made. The FIB uses a beam of ions to remove material and then thin an electron transparent region of interest directly from a wafer to allow characterization in the TEM. In this chapter, we will briefly discuss the operation of the FIB, then ion beam–sample interactions and specimen damage, followed by a detailed explanation on how to make a TEM sample in the FIB. The chapter ends with some brief discussion of conventional Ar ion milling and wedge polishing.

Chapter written by David COOPER and Gérard BEN ASSAYAG.

9.1. The focused ion beam tool

The FIB typically uses a 30-kV Ga ion beam to sputter material from a sample and to produce an electron transparent lamella whose thickness can be controlled to an accuracy of some tens of nanometers. Secondary electrons generated from the milling process can be used to form an image that enables the Ga ion beam to be positioned accurately on the sample. The Ga ions in the FIB are emitted from a liquid-metal ion source (LMIS) at high vacuum. Ga is liquid just above the room temperature and flows into a tungsten needle of radius 2–5 μm. An electric field (10^{10} Vcm^{-2}) is applied to the needle, which causes the liquid Ga to form a source of around 2–5 nm in diameter in the shape of a Taylor cone [TAY 64]. The apex of the Taylor cone, formed by the balance between the Ga surface tension and the electrostatic forces, is small enough, and the local electric field is high enough to field evaporate and simultaneously ionize the Ga from it, resulting in a Ga$^+$ ion beam. The amount of Ga ions that is extracted is replenished by the flow of liquid Ga into the Taylor cone. The required characteristics of the liquid metal used in the LMIS are a low melting point, low vapor pressure and the ability of the metal to wet the needle; Ga fulfills these criteria. Alloy sources can be used, but these introduce complications regarding stoichiometry and the overall performance of the FIB system.

Once Ga ions are extracted from the LMIS, they are accelerated down the ion column, using a typical voltage range of 30 to 2 kV. The ions pass through a condenser lens and apertures to define the probe size, and an objective lens is used to focus the ions. Several apertures are used to control and optimize the probe size for a given beam current. The current density can be altered from a few pA to several nA, corresponding to beam diameters of 5 nm–0.5 μm. The highest current achievable is about 10 Acm^{-2}. The ion beam distribution can be described as having a Gaussian profile with long "tails". Therefore, although the Ga ion beam has a strong central spot with a high current density suitable for accurate sputtering, the beam tails will affect regions close to the sputtered regions, removing material at a lower rate and implanting surfaces with Ga. Thus a low current is used for accurate milling, when minimum Ga ion implantation is required.

The imaging mechanism in the FIB involves the detection of secondary electrons that are ejected from the specimen during ion bombardment. Although it is possible to form an image using the secondary ions that are sputtered from the sample, thus acquiring complementary elemental information, imaging using the electrons is normally preferred as the yield is approximately 1,000 times higher [GIA 04]. Figure 9.1 shows a schematic diagram of a single Ga ion interaction in a target sample. Secondary electrons are detected as the ion beam is rastered over the sample surface, forming an image as for standard SEM. The imaging is affected by specimen charging because non-conducting regions accumulate positive charge, preventing the secondary electrons from escaping. Apart from reduced imaging capabilities caused by the charging, a more critical effect is ion beam deflection as the charge builds up. Sample charging can be reduced by coating the specimen with gold or carbon; another solution is to place the needle used for *in situ* lift-out onto the specimen surface in order to remove the charge.

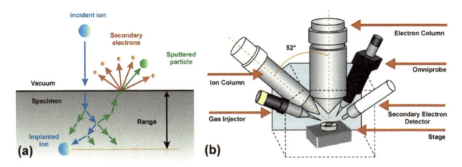

Figure 9.1. *(a) Schematic diagram of the FIB sputtering process and ion-solid interactions. (b) Schematic of the arrangement inside a modern dual-beam FIB*

Most modern FIB systems are known as dual beams as they use a separate electron column for imaging in combination with the Ga FIB column that is used for milling. The dual beam system is attractive, as the specimen can be observed without implanting Ga ions unnecessarily for imaging. Figure 9.1(b) shows a schematic of the interior of a modern dual-beam FIB tool.

9.2. Ion-sample interaction

The FIB provides an easy and reliable means to prepare specimens containing individual devices with nm-scale precision. However, it is important to fully understand what happens to the specimen during preparation. The FIB operates by sputtering material from the specimen using kinetic energy transfer from incident high-energy Ga ions. Displacement of atoms from their equilibrium lattice positions results from either a single collision with the ion or more generally a collision cascade in the material [GIA 04]. The sputtering yield is defined as the number of atoms ejected per projectile and is dependent on the incident angle and energy of the ion. Figure 9.2(a) shows the angular dependence of the sputtering yield for 30 kV Ga and 3 kV Ar ions in amorphous silicon, calculated using "Stopping Range of Ions in Materials" (SRIM) simulations [ZIE 03]. SRIM simulations assume an amorphous target and, therefore, do not account for effects such as channeling. Nonetheless, they do provide an insight into the comparative effects of ion-sample interactions.

At normal incidence, most Ga ions are implanted deep into the specimen. The probability of a Si atom being ejected from the specimen surface is low. At shallower angles, the sputtering yield increases until a maximum is reached, before decreasing at glancing angles. For an atom to be sputtered from the lattice, the energy transfer from an incident particle must be much greater than the lattice surface energy. The surface binding energy of a specimen defines the sputtering rate of a material. Figure 9.2(b) shows the surface energies for different target materials as a function of Z [ZIE 03]. The semiconductors Si and Ge are marked, showing a relatively high surface binding energy. Zn, for instance, has a very low surface binding energy and is known to sputter rapidly [GIA 04].

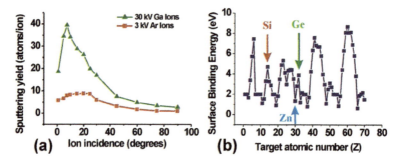

Figure 9.2. *(a) Sputtering yields for 30-kV Ga and 3-kV Ar ions incident on amorphous Si calculated using SRIM simulations where 90° represents normal incidence. (b) Surface binding energies for amorphous target elements as a function of Z, calculated using SRIM*

During ion milling, a dual process of implantation and sputtering occurs. After an amount of material has been sputtered that is comparable to the projected range of the ion in the material, a steady-state Ga concentration is reached [ISH 98]. The range is defined as the penetration depth of an ion in the target material. The limit of the Ga implantation is proportional to the reciprocal of the sputtering yield, such that a material that sputters quickly will have a lower Ga ion concentration implanted than a material with a low sputtering rate. Care must be taken when observing polycrystalline specimens prepared using FIB milling as the Ga ions can diffuse to the grain boundaries and modify their properties [UNO 10].

Irradiation of a target with a beam of ions will result in microstructural disruption to the lattice. For most materials, there will be a critical implantation dose above which the lattice disruption is so acute that the material is amorphized. A sidewall layer that has been exposed to Ga ions at glancing angles in a Si specimen prepared using a 30 kV Ga ion beam typically has an amorphous layer thickness of around 25 nm, which can be reduced by decreasing the ion energy. In regions where the ion dose is beneath the critical implantation dose, extensive defect creation will occur, such as the creation of vacancy-interstitial pairs. Table 9.1 summarizes the extent of the damage introduced in a target material as a function of ion dose [WIL 98]. Here, it is important to consider that although there is the presence of the frequently discussed amorphous layer, specimen preparation by FIB milling will also introduce defects deep in the specimens that are less obvious to observe and may also modify the measured properties. This particular problem is especially known for dopant potential measurements by off-axis electron holography.

Dose	Cascade process	Structural effects
10^8 cm^{-2} (ppb)	Individual tracks	Discrete defects
10^{12} cm^{-2} (ppm)	Overlapping tracks	Complex defects
10^{15} cm^{-2} (1%)	Complete overlap	Amorphization
10^{18} cm^{-2} (50%)	Composition changes	Compound formation

Table 9.1. *Summary of target damage mechanisms for different ion doses [WIL 98]*

Figure 9.3(a) shows the range of 30-kV Ga and 3-kV Ar ions in amorphous Si, calculated using SRIM simulations as a function of incident angle, with 90° being normal to the specimen surface. To provide a Si sample using a 30-kV ion beam with parallel sides, an angle of 1° is required. At this angle, the simulations suggest that the average and maximum Ga ion range in a Si substrate is 10 and 15 nm, respectively. The simulated implantation ranges are less than the experimentally observed values of 25 nm; however, they can give an estimation of the relative amounts of damage observed from different types of ions at different energies in different types of materials. Figure 9.3(b) shows a high-resolution TEM image of a typical Si specimen that has been prepared using an FIB at 30 kV. The crystalline and amorphous layers appear to have an abrupt junction. However, the presence of damaged regions in the crystalline regions can also be clearly seen by the dark contrast.

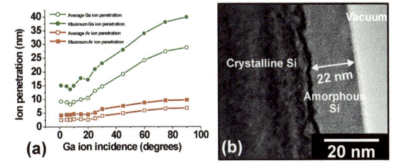

Figure 9.3. *(a) 30 kV Ga and 3 kV Ar ion penetration ranges generated using SRIM and plotted as a function of angle of incidence in amorphous Si. Mean and maximum penetration depths are shown. (b) High-resolution TEM image of 250-nm-thick Si specimen after Ga ionbeam milling at 30 kV showing the damaged crystalline and amorphous regions*

Figure 9.4 shows SRIM simulations of different ions incident onto amorphous Si. Here, an angle of 1° has been used for Ga ion FIB milling and 5° has been used for Ar and Xe. The paths indicate the trajectories of the incident ions and subsequent displacements caused by cascades. Each of the atoms that are displaced from their lattice site then creates a significant number of additional defects in the lattice. The SRIM simulations do not account for the effects of channeling that occur when incident ions are aligned with major crystallographic directions or for the vacancy-interstitial recombination due to temperature. For channeled ions, the principal energy loss mechanism is from interactions with the electrons in the lattice. As a result, some incident ions can have a significantly longer range than for amorphous materials [STE 90]. Besides channeling from incident Ga ions in the FIB, the lattice ions displaced during the cascade process may also be steered into ion channels, resulting in a deep disorder in the specimen.

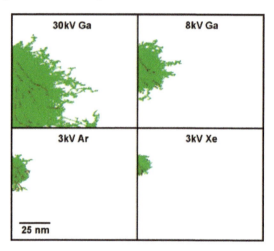

Figure 9.4. *Simulations of different ions in amorphous silicon at typical angles of incidence for FIB and broad beam milling (the ions enter the target from the top and left, Ga ions at 1°, Ar and Xe ions at 5°). The energy and the size of the ions strongly affect the range of the ions in the material*

The simulations suggest that heavy ions at low energies and angles of incidence are best for specimen preparation. Light ions will have a longer implantation range inside the specimen, whereas heavy ions will stay close to the surface and be sputtered away. Ar is from the point of view of specimen damage, worse than Ga. However, typically Ar milling is performed at a lower energy (although with a higher angle of incidence) than FIB milling with Ga ions. Xe are perfect ions for specimen preparation, being relatively heavy and also compatible with conventional broad beam ion millers [CHE 87].

During ion milling, most of the sputtered material is rapidly pumped away into the vacuum system. However, some sputtered atoms may redeposit on the freshly milled walls of the specimen, resulting in the formation of an additional partially crystallized, rough surface layer. Therefore, for final milling steps, it is best to alternate the milling between the faces of the specimens to avoid re-deposition in the region of interest.

9.3. Beam currents and energies for specimen preparation

Generally, the aim for specimen preparation is to provide a parallel-sided membrane containing a region of interest with minimum surface damage. For most analysis such as high-resolution imaging, quantitative electron energy loss spectroscopy (EELS) or energy dispersive X-ray spectroscopy (EDX) and strain mapping, a good guide is to prepare specimens that are thinner than the mean free path of an electron in that material; this is often 100 nm or less. For most quantitative TEM techniques, the thickness of the specimen needs to be known. If the specimen is wedge shaped, then the thickness of the specimen needs to be locally mapped that will introduce errors into the analysis and may also compromise the accuracy of the experiment. Using an FIB, a specimen can be produced with only a few nanometers of thickness variations across a field of a micron or more. For practical specimen preparation, the shape of the beam at different currents and the quality of the ion beam focusing at low energies need to be considered in order to achieve thin and perfectly parallel-sided specimens.

Figure 9.5 shows line scans cut into a Si substrate using a FIB operated at 30 kV for different beam currents. The measured width of the cut at the specimen surface is also shown, compared to the nominal beam diameter shown in brackets. The cuts demonstrate that the ion beam has a Gaussian form and the tails of the beam lead to a width of the cut that is significantly larger than the nominal diameter. To avoid the effects of the tails of the beam, low currents need to be used in combinations with slight specimen tilts in order to achieve specimens with parallel sides. Of the beam currents used here, 22 nA is considered very high and is only used for rough milling; the current of 2.8 nA is used for rough milling near the sample. Medium currents between 100 and 280 pA are used for metal deposition and finer milling. Specimens are generally finished using a lower current of between 20 and 100 pA. The beam current used for the final milling steps can be significantly higher with recently developed machines, but the principles remain the same.

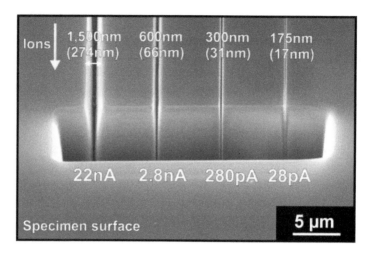

Figure 9.5. *Line scans cut into a Si substrate using 30-kV Ga ions at different currents*

Figure 9.6. *Ion imaging of 500-nm-thick SiO layers on Si for different FIB operating voltages*

Figure 9.6 shows beam images of 500-nm-thick polysilicon layers on Si acquired using ion beam imaging at different FIB operating voltages using a first-generation dual-beam FEI Strata 400 system. At 30 kV, it is easy to obtain good quality images because the ion beam is well focused. However, at voltages below 8 kV and certainly below 2 kV, the ion column struggles to finely focus the beam and this is reflected in the poor quality images. As a result, it is difficult to both visualize the sample and provide precise milling. For good control of the sample and a reduction of the damage that is introduced, an operating voltage in the range 8–5 kV is a good compromise. Lower voltages can be used although these may introduce a wedge shape to the specimen. For very low energies, care must be taken to ensure that the Ga ions actually perform additional milling and are not just implanted into the specimen surfaces.

Either increasing the beam current or decreasing the operating voltage in the FIB will have the effect of increasing the size of the FIB that is used to thin the specimen. As a guide, Figure 9.7 shows the nominal beam spot size for different beam energies and currents for a FEI Strata 400 FIB Dual Beam System.

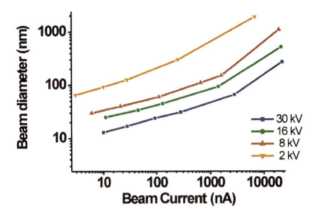

Figure 9.7. *Nominal beam diameters as a function of the beam current for different operating voltages in an FEI Strata 400 FIB Dual Beam System*

Ideally, all specimens would be finished using the lowest beam currents and the lowest operating voltages; however, if the milling rate is too low, then specimen drift caused either from physical stage movement or from image drift caused by charging will limit the amount of time that can be used to performs the final milling steps on the specimens. For practical specimen preparation, a compromise needs to be made between beam energy (damage reduction and spot size), current (spot size) and milling time (drift).

Combinations of the sputtering rate of the material being examined, the beam energy and current will result in a specimen side with different angles relative to the ion beam. To provide a parallel-sided specimen, due to the Gaussian shape of the beam, the specimen faces need to be tilted into the ion beam. As an example, when using low currents and an operating voltage of 8 or 2 kV to mill silicon, angles of 2° and 7° are used, respectively. For materials with a higher milling rate, such as InP and some metals, such as Zn, a much higher angle of incidence, than for Si, is required. For different materials, the thickness gradient across the specimens for different angles of incidence should be mapped using EELS techniques or electron holography in order to assess the best milling conditions for providing parallel faces.

As previously discussed, damage may be found deep in the crystalline regions of the specimens and it is not always enough only to think of the amorphous layers when using low operating voltages to clean the specimens. Depending on the information that is needed from the experiment, the author will typically reduce the milling voltage from 30 to 8 kV when the specimen is more than 1 µ thick, so that several hundreds of nanometers are removed from each side using the lower beam energy. Similar considerations should be used when using lower FIB operating voltages of additional Ar milling for finishing.

9.4. Practical specimen preparation

There are many variants of FIB specimen preparation; here we focus on two of the most commonly used methods *in situ lift-out* and *H-Bar*. There is no fixed method of preparing specimens; the choice of beam energies and currents depends on the types of samples being examined and the depth of the region of interest. This section only provides a guide to emphasize the most important steps. The best way to make good specimens is by practice, trial and error.

9.5. *In situ* lift-out

TEM sample preparation using lift-out was first developed to improve the site specificity of FIB-prepared membranes [OVE 90]. Figure 9.8 shows a schematic of the basic lift-out procedure for a sample. Material is milled from around the region of interest by cutting some steps using a high beam current; then the specimen is tilted so that the base of the region of interest can be cut away before it is extracted from the wafer and stuck to a TEM grid. There are many different ways of performing *in situ* lift-out. Here, we will discuss only the step method. The advantage of using this method is that the user can always see the depth of the milled regions. This is very useful when performing lift-out on materials that are very tough such as sapphire and silicon carbide, or when working with new materials where the milling rate is unknown. The disadvantage of this method is that it is relatively time consuming when compared to methods such as chunk extraction.

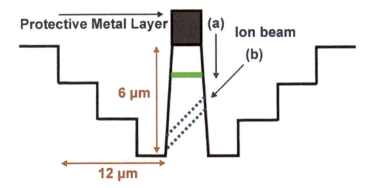

Figure 9.8. *Schematic diagram illustrating the principal for in-situ lift out. The region of interest is cut from the wafer by sputtering material from the sides using the ion beam (a) at normal incidence and then (b) from an angle of incidence of 45°*

Figure 9.9 shows the initial steps used for *in situ* lift-out on a fully processed semiconductor wafer: (a) the region of interest is located using the electron beam; and (b) a thick protective strip of Pt or W is deposited above the region of interest using ion beam-assisted deposition (IBAD) [KAI 90]. This step is important for locating the region of interest when ion imaging, as for insulating semiconductor wafer surfaces often nothing can be seen when using ion imaging; it also protects the region of interest from normal incidence ions. However, it is important to deposit several microns of metal onto the specimen surface if very thin specimens are required. The tails of the ion beam will mill the deposited metal during thinning and when all of the metals have gone; it is not possible to do any more milling. If a specimen thickness of 100 nm is targeted then around 2 µm of metal deposition is required. For a thinner specimen, it is a good idea to put more. As metal deposition uses the ion beam to collide with a gas flow to grow the metal layer, then it is clear that this stage will damage the region directly below the specimen. If the region of interest is close to the surface, then electron beam-assisted deposition (EBAD) can be used to protect this top layer [KAI 90]. Other approaches such as sputtering a metal coating onto the specimen surfaces can be considered. Figure 9.9(b) shows an electron image viewed from an angle of 52°, the material on either side of the region of interest has been removed using step milling at a high beam current. The length of the milled boxes needs to be approximately twice the depth of the milling in order to see the base of the specimen when it is tilted to provide the cuts for lift-out. The current is then reduced to around 2.8 nA and the specimen is thinned to approximately 2 µm as shown in Figure 9.9(c). To free the specimen for lift-out, the stage is tilted to 45° relative to the ion beam and cuts are made around the region of interest as shown by the ion image in Figure 9.9(d). Note the lack of contrast on the insulating wafer surface. A small amount of material is left at one end of the sample to hold it in place. For this step, parallel milling is required, sequential milling of the three different boxes will result in redeposition of material in the cuts. The undercuts should be performed well away from the region of interest to avoid possible damage. Figures 9.9(e) and (f) show electron images of the specimen after the undercuts have been made with the beam at 0° and 45° incidence, respectively. It is clear from the dark contrast around the region of interest when the specimen has successfully been freed from the substrate.

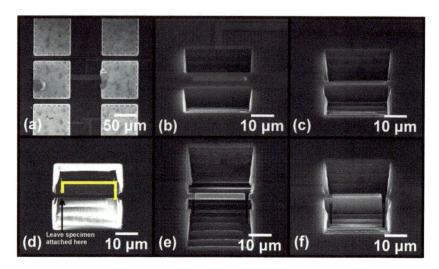

Figure 9.9. *Shows the preparation steps for in-situ lift out*

There are many complicated methods for performing the *in situ* lift-out of TEM specimens. However, after some practice, a system such as the Omniprobe that uses a tungsten needle is very reliable and easy to use. The cost of only a few euros for each needle makes the inevitable accidents that do occur during the learning process less stressful. Figure 9.10(a) shows the micromanipulator and also the gas deposition system in the FIB chamber. The micromanipulator is carefully brought to the lamella. Figure 9.10(b) shows micromanipulator tungsten needle at high magnification just before contact with the sample, the needle can clearly be seen to vibrate (Figure 9.10(c)) as soon as the needle makes contact with the sample, the vibration stops and a change in the image contrast can be seen. Figure 9.10(d) shows how small an amount of metal is used to stick the tip apex to the surface of the sample. Typically a 1 μm layer over an area of 1 μm^2 is enough that takes only a few seconds to deposit using a current of 0.28 nA. Finally (Figure 9.10(e)), a cut is made to free the sample and (Figure 9.10(f)) the sample is removed from the wafer.

Specimen Preparation for Semiconductor Analysis 231

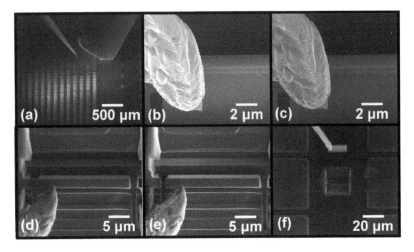

Figure 9.10. *The extraction steps used for in-situ lift out*

Figure 9.11 shows SEM images illustrating the steps for welding the specimen to the grid and final milling. Figure 9.11(a) shows the needle approaching the grid. As the gas injector system is often located on the opposite side of the FIB chamber when compared to the micromanipulator, care must be taken to ensure that there is enough gas flow across the region of interest to successfully stick the sample to the grid. Figure 9.11(b) shows how the specimen is placed near the grid, leaving enough space for the metal deposition to securely stick the sample. Figure 9.11(c) shows that once the specimen is securely attached to the grid then the probe can be cut from the sample and carefully removed from the chamber.

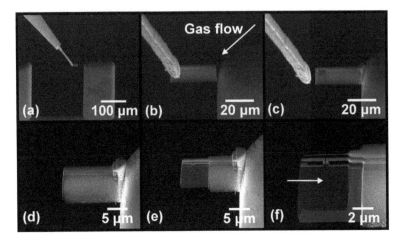

Figure 9.11. *The steps used for welding the sample to the grid and final thinning*

Figures 9.11(d) and (e) show the final thinning steps, first the end of the sample is cut away so that the region of interest is a few microns from the specimen edge as long samples are prone to bending. At this higher magnification, redeposition from the rough milling that was used before lift-out can now be seen on the specimen surfaces. Continuing to use an operating voltage of 30 kV but with an intermediate beam current of around 0.28 nA, the specimen is thinned to around 1.0 μm using cleaning cross-sections. In each case, the specimen is tilted into the ion beam by a degree in order to achieve parallel sides. Once the sample has been thinned to around 1 μ, the beam energy can be reduced to a lower voltage for the finishing steps and the specimen can be thinned by alternating cleaning cross-sections on either side until the final dimensions are reached. Here, to keep the parallel sides at these lower beam energies, the specimen is now tilted into the beam by 2.0°. Figure 9.11(f) shows the device in the sample finished using an ion beam energy of 5 kV and a current of 66 pA. The metal deposition on the surface has almost been entirely removed and no further milling is now possible without exposing the region of interest to the tails of the beam. If a very large region of the specimen needs to be thinned, then it may be faster to step down the operating voltage of the FIB gradually using intermediate steps such as 30, 16, 8 and then 5 kV.

Note the thickness variations indicated by the arrow located under the metal plugs in Figure 9.11(f). This is known as curtaining and is due to differential milling rates between the metal and the oxide regions above the specimen. For some experiments, these thickness variations can make analysis difficult when they appear through a region of interest. A solution to this is to use back-side milling where the specimen is rotated by 180° and then thinned from the silicon substrate side where no differential milling can take place. Although methods to do this are commercially available, it is reasonably straightforward to find a homemade solution where the specimen is stuck to the end of a needle, which is then manually rotated outside of the FIB chamber.

9.6. H-bar technique

The H-bar technique is the most straightforward method of FIB sample preparation for TEM analysis and can be performed without either the use of an micromanipulator or a SEM [PAR 90]. A slice of material is mechanically polished to a thickness of between 20 and 100 μm and glued to half of a Cu TEM grid using conducting silver epoxy. The sample is then mounted in a dedicated TEM preparation clamp and transferred to the FIB.

Figure 9.12 shows a series of SEM images of a specimen, acquired in the FIB. Figure 9.12(a) shows the deposition of a protective metal layer over the region of interest and initial rough milling, performed at a high beam current. If a very thin specimen has been mechanically polished to a thickness of less than 50 μm, then this rough milling stage can be as short as 15 min. The polished specimen should be robust enough to survive further handling and for Si the optimal thickness is around 50 μm. Figure 9.12(b) shows a finished specimen that has been thinned at progressively lower beam currents and energies. Five membranes with different thicknesses in the range of 200–500 nm are shown in Figure 9.12(b). Figure 9.12(c) shows the sample tilted to 45° to illustrate the technique. Redeposition can be seen in the form of small round structures around the lamella. Other than a high success rate and simplicity, the H-bar technique offers few advantages when compared to *in situ* lift-out and makes site specificity challenging. However, one advantage is that it provides relatively large areas where electrical or thermal contacts can be placed close to the region of interest for *in situ* biasing, straining or heating experiments.

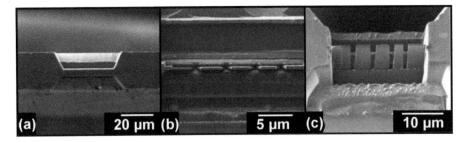

Figure 9.12. *The steps used for H-Bar FIB milling*

9.7. Broad beam ion milling

Broad beam ion milling can be used to prepare specimens by itself or to clean FIB-prepared specimens. Figure 9.13(a) shows low magnification and (b) high-resolution images of a specimen prepared in the FIB operated at 30 kV. The high-resolution image in Figure 9.13(c) shows that the amorphous layer has been reduced to less than 5 nm after using only low-energy, low-angle ion milling Ar ions. Although it is extremely easy to thin FIB-prepared specimens using Ar milling, care must be taken to avoid surface roughening and redeposition from the copper grid on the regions of interest. This can be avoided by carefully choosing the trajectory of the ion beams relative to the specimen.

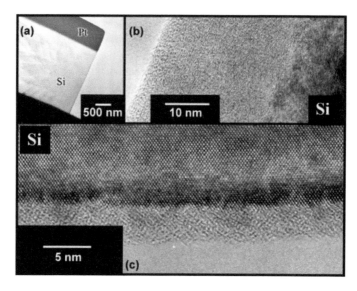

Figure 9.13. *(a) FIB-prepared specimen at low magnification. (b) An amorphous layer of 25 nm has been measured by HRTEM and (c) this is reduced to less than 5 nm after low-energy Ar milling*

Specimens can also be prepared using "conventional Ar ion milling" that uses a combination of mechanical polishing and low-energy Ar ions to prepare electron transparent specimens suitable for analysis by TEM. Typically, a specimen is mechanically thinned to around 20 µm and a hemisphere is then ground into the underside, leaving a thickness of less than 2 µm. The specimen is then milled with low-energy, low-angle Ar ions that sputter material from the sample until a small hole appears in the thinnest area. The regions near the milled hole will typically be 10–100 nm in thickness that is optimal for many TEM applications. Conventional Ar ion milling is an ideal technique for applications that require a large, thin region for characterization.

There are numerous disadvantages of using this technique, such as the presence of non-uniform thickness variations, due in part to differential milling rates between different materials [UNO 10, WIL 98], and its non-site specificity. More critically, as with all ion beam sputtering techniques, the surface regions of the sample are damaged, leading to an amorphous surface layer and significant Ar ion implantation

9.8. Mechanical wedge polishing

Mechanical polishing involves the use of diamond lapping films to thin a specimen mounted on a glass stub, successively using finer diamond particle sizes until the sample has been finished using a 0.1 µm film. The sample is then turned over and tilted to an angle of between 5° and 1° and polished. Specimens can be finished using a 0.02 µm colloidal silica suspension on a cloth polishing mat. This method of sample preparation is an art and requires practice. This method has the additional benefit that ion milling is not obligatory, but in practice some additional ion milling is normally used on mechanically polished specimens to additionally thin the sample and to remove debris.

The disadvantages of this method of sample preparation are numerous. Often, significant amounts of debris, from the sample, or from the wax used to secure the sample to the mounting stub can be deposited on its surfaces. Even though samples may look very smooth in an optical microscope, the higher magnification available in the TEM reveals that the surfaces may be rough. Wedge-polished specimens are very fragile, as the thin areas are neither protected nor supported and can easily break. Finally, as with cleaving, mechanical polishing is only site specific to a few hundred microns, not fulfilling the requirements for modern device characterization. However, if good quality specimens can be prepared using wedge polishing, for example for the examination of blanket films where site specificity is not required, then this remains a wonderful method for providing relatively undamaged materials for observation.

9.9. Conclusion

Specimen preparation is the most important stage for TEM analysis. A good sample will make the TEM experiments and subsequent analysis of the data much easier. For semiconductor research, the FIB tool is an extremely powerful method of preparing specimens for TEM observation with nm-scale site specificity. Combinations of FIB milling, mechanical polishing and broad beam milling using Ar or Xe ions can be used to provide flexible preparation solutions for different types of materials. However, besides mastering the control of the FIB, it is important to understand the physics of ion-sample interaction in order to understand the extent of the damage that is introduced into the specimen. The modified regions in a specimen can be deep in the crystalline regions and it is often not enough just to clean off the amorphous layers using low energy milling. For rough observation of structures by TEM, sophisticated milling recipes are not always required, but for the quantitative analysis of composition, dopant potential or strain, the experiment can only be performed with an appropriate, well-prepared sample. The preparation of the new materials and systems that are encountered in the semiconductor research is often complicated, but by mastering the steps that are required, the microscopist has the freedom to iterate and improve their methods in order to provide the perfect specimens for observation. Further improvements in specimen preparation are required in terms of resolution, efficiency and throughput. Higher resolution at very low voltage is desirable and even if much progress has been made recently, much remains to be done.

9.10. Bibliography

[CHE 87] CHEW N.G., CULLIS A.G., "The preparation of transmission electron microscope specimens from compound semiconductors by ion milling", *Ultramicroscopy*, vol. 23, pp. 175–198, 1987.

[GIA 04] GIANNUZZI L.A., STEVIE F.A., *Introduction for Focused Ion Beams*, Springer, New York, 2004.

[ISH 98] ISHITANI T., KOIKE H., YAGUCHI T., KAMINO T., "Implanted gallium ion concentrations of foused ion beam prepared cross-sections", *Journal of Vacuum Science & Technology B*, vol. 16, pp. 1907–1913, 1998.

[KAI 00] KAITO T., FUJITA J., KOMURO M., KANDA K., HARUYAMA Y., "Three-dimensional nanostructure fabrication by focused ion beam chemical vapour deposition", *Journal of Vacuum Science & Technology B*, vol. 18, pp. 3181–3184, 2000.

[OVE 93] OVERWIJK M.J.F., "A novel scheme for the preparation of transmission electron microscopy specimens with a focused ion beam", *Journal of Vacuum Science & Technology B*, vol. 11, pp. 2021–2024, 1993.

[PAR 90] PARK K.H., "Cross-sectional TEM specimen preparation of semiconductor devices by focused ion beam etching", *Materials Research Society Symposium Proceedings*, vol. 199, pp. 271–280, 1990.

[STE 90] STECKL A.J., MOGUL H.C., MOGREN S., "Ultrashallow Si Pn junction fabrication by low energy gallium focused ion beam implantation", *Journal of Vacuum Science & Technology B*, vol. 8, pp. 1937–1940, 1990.

[TAY 64] TAYLOR G.I., "Disintegration of water drops in an electric field", *Proceedings of the Royal Society of London A*, vol. 280, pp. 383–397, 1964.

[UNO 10] UNOCIC K.A., MILLS M.J., DAEHN G.S., "Effect of gallium focused ion beam milling on preparation of aluminium thin foils", *Journal of Microscopy*, vol. 240, pp. 227–238, 2010.

[WIL 98] WILLIAMS J.S., "Ion implantation of semiconductors", *Materials Science A*, vol. 253, pp. 8–15, 1998.

[ZIE 03] ZIEGLER J.F., ZIEGLER M.D., BIERSACK J.P., *Nuclear Instruments & Methods B*, vol. 268, pp. 1818–1823, 2003.

List of Authors

Gérard BEN ASSAYAG
CEMES/CNRS
Toulouse
France

Patrick BENZO
STMicroelectronics/CEMES
Grenoble
France

Nikolay CHERKASHIN
CEMES/CNRS
Toulouse
France

Alain CLAVERIE
CEMES/CNRS
Toulouse
France

Laurent CLÉMENT
STMicroelectronics
Crolles
France

David COOPER
CEA-LETI, Minatec
Grenoble
France

Dominique DELILLE
FEI Company
Eindhorer/Acht
Netherlands

Christophe GATEL
CEMES/CNRS
Toulouse
France

Florent HOUDELLIER
CEMES
Toulouse
France

Martin HŸTCH
CEMES/CNRS
Toulouse
France

Elsa JARON
CEMES
Toulouse
France

Aurélien MASSEBOEUF
CEMES/CNRS
Toulouse
France

Roland PANTEL
STMicroelectronics
Crolles
France

Shay REBOH
CEA/LETI
Grenoble
France

Sylvie SCHAMM-CHARDON
CEMES/CNRS
Toulouse
France

Germain SERVANTON
STMicroelectronics
Crolles
France

Etienne SNOECK
CEMES/CNRS
Toulouse
France

Index

113 defects, 185
2D dopant distribution, 54

A

amorphous layer, 222, 223, 228, 234
analytical transmission electron microscopy, 135
annular dark field (ADF), 137
arsenic, 41
atomic structure, 139-140

B

bending, 82
Bohr magneton, 126
bubbles, 187-189
burgers vector, 165-193

C

chemical reaction, 135-158
CMOS and BiCMOS transistors, 50
CNT/Carbon Nanotube, 205
contact etch stop layers (CESL), 96
contacting, 88
convergent beam electron diffraction (CBED), 65
crystallization state, 138
current injection, 200, 204, 209

D

Dark Field Electron Holography (DFEH), 167
Dark-field electron holography (DFEH), 81
defects, 165-193
device, 199
dielectric constant, 136
direct wafer bonding (DWB), 167-170
dislocations, 166-172
Dopant profiling, 1-33
Dopant segregation, 38
dual-beam, 221, 226

E

electro-magnetic field, 200
electron beam induced deposition (EBID)
electron diffractionm, 65-77
electron energy loss spectroscopy (EELS), 41, 135
electron holography (EH), 99, 107-132, 223, 227
electron wave, 107
electrostatic phase shift, 109
elemental profile, 145, 151, 153

energy dispersive X-ray spectroscopy (X-EDS), 46-49
energy filter, 142, 143
EOR defects, 173

F

FIB/Focused Ion Beam, 208-209
FinFET, 101-102
finite element modeling, 67
focused ion beam (FIB), 219
focused ion beam milling, 14
fresnel contrast, 165, 189

G

g.b analysis, 168
gas injection system (GIS)
gas platelets, 189-193
gate oxide, 135, 138, 142, 144
Ge, 144-155
geometric phase analysis (GPA), 82, 165
grain boundary, 51, 55, 56

H

habit plane, 177-193
H-Bar lamella, 233
heating, 201
high annular dark field (HAADF), 137
high resolution transmission electron microscopy (HRTEM), 137
High-k, 135
high-resolution transmission electron microscopy (HRTEM), 82
HoloDark, 88
hydrogen implantation, 93

I, K

inactive thickness, 30-33
inside/outside contrast, 181-185

interdiffusion, 135-158
interfacial layerm, 136
interferometry, 108
ion beam induced defects, 223
ion beam induced deposition (IBID)
ion beam milling, 229
ion bombardment, 220
ion implantation, 172-173
irradiation, 201, 202, 206
kinematical and dynamical simulations, 68

L

liquid metal ion source (LMIS), 220
Lorentz microscopy, 87, 110

M

magnetic devices, 126
magnetic fields, 125, 126
magnetic induction, 108, 110
magnetic phase shift, 109
magnetization recording, 108
mean inner potential (MIP), 4
MEMS, 200
mobility, 65, 72, 77
MOS transistors, 72
MOSFET, 135, 139, 140
movable probe, 202
multi-contact, 204-205
multiple linear least square (MMLS), 151

N, O

nano beam diffraction (NBD), 65
nanomagnetism, 121
nanomaterials, 107, 113
nitride membranes/TEM window, 209-210
n-MOSFET, 94-96
off-axis electron holography, 1-33

P, Q

permittivity, 138, 144
perpendicular magnetic recording, 108
phase image, 3-15
phosphorus, 41
p-MOSFET, 87-92
p-n junctions, 40, 54
precipitates, 187
quantification/quantitative, 200

R

radiation damages, 43-44
rare earth oxide, 135
recessed sources and drains, 87-92
relaxation, 66

S

sample holder, 200-210
scanning, 142, 145
Schottky Field Emission Gun, 142
semiconductor characterisation, 2
Si, 144-155
SiGe heterostructures, 170-172
silicate, 136, 142, 156
silicon doping, 37-38

silicon drift EDS detector (SDD), 41
SiO_2, 135
SOI, 101
specimen preparation, 28-29
spectrum imaging, 49
spectrum quantification, 47
spherical aberration correction, 137, 145,
spintronic, 107
statistical analysis, 185-187
strain mapping, 81-102
strain, 65, 88, 166
strained silicon, 87-92
stray scattering, 149
stress, 65, 81-102
structural and chemical thickness, 151-155

T

thin-film relaxation, 92
transistor, 205
transition metal oxide, 144

V, W

voids, 187-189
weak beam dark field (WBDF), 165